Foraminifera

Alcide d'Orbigny, from a lithograph by Lavallée. Published as a frontispiece in "Voyage dans l'Amérique Méridionale" Vol. 1.

Foraminifera

Edited by

R H HEDLEY

British Museum (Natural History)
London

C G ADAMS

British Museum (Natural History)
London

Volume I

1974

ACADEMIC PRESS London and New York
A Subsidiary of Harcourt Brace Jovanovich, Publishers

ACADEMIC PRESS INC. (LONDON) LTD.
24/28 Oval Road,
London NW1

United States Edition published by
ACADEMIC PRESS INC.
111 Fifth Avenue
New York, New York 10003

Library of Congress Catalog Card Number: 74–5662
ISBN: 0 12 336401 9

Text set in 11/12 pt. Monotype Baskerville, printed by letterpress,
and bound in Great Britain at The Pitman Press, Bath

Contributors

ZACH M. ARNOLD, *Department of Paleontology, University of California, Berkeley, California 94720, U.S.A.*

Y. LE CALVEZ, *École Practique des Hautes Études, Laboratoire de Micropaleontologie, 8 rue de Buffon, Paris Ve, France*

JOHN J. LEE, *Department of Biology, The City College of New York, Convent Avenue and 138 Street, New York NY 10031, U.S.A.*

ALFRED R. LOEBLICH JR., *Department of Geology, University of California, Los Angeles, California 90024, U.S.A. and Chevron Oil Field Research Company, La Habra, California 90631, U.S.A.*

G. H. SCOTT, *New Zealand Geological Survey, PO Box 30, 368 Lower Hutt, New Zealand*

HELEN TAPPAN, *Department of Geology, University of California, Los Angeles, California 90024, U.S.A.*

Preface

Why another work on Foraminifera? After all, many books have already been written on this group of protozoans and several journals exist, of which no less than four are devoted solely to micropalaeontology. Can there be room, not only for another book, but for the three volumes planned in this series? Is there a need at the present time for further synthesis? We believe that there is and that it stems from the dichotomous growth of foraminiferal studies.

For more than a hundred and fifty years zoologists and palaeontologists have shared in the study of foraminifera, their common interest originally being the taxonomy of the group. By the turn of the century, however, zoologists were beginning to explore more thoroughly such biological aspects as life histories, cytology and, more recently, ecology and nutrition. Palaeontologists, meanwhile, took another route. With the rapid expansion of the oil industry, they found themselves hard pressed to supply the refined taxonomic and biostratigraphical data on which detailed correlation could be based. But this economic pressure gave direction to their work, with the consequence that evolutionary knowledge grew apace while systematics made giant strides.

Latterly, palaeontologists have broadened their interests to include ecology (as a means of interpreting palaeoecology) and also cytology. In this way they have entered fields previously dominated by zoologists. The two groups, each with its own methodology, have rather different approaches to common problems, for example, shell structure, ecology, and the recognition of species limits in space and time. Misunderstandings have arisen, and work has tended to be duplicated or to be undertaken in partial ignorance of earlier advances. This is readily understandable, and in fields such as biometry, biogeography, evolution and zonation, the volume of data is now so large and so widely dispersed throughout the biological and palaeontological literature that there is urgent need for collation and appraisal.

By inviting stimulating surveys of progress by specialists in various fields we hope to promote a better understanding of the branches that now contribute to foraminiferal research. The series is also intended as a guide to recent literature, both for the student and the specialist.

The invitations already accepted should provide material for further volumes and we hope that subsequent papers will enable the series to be continued. Subjects planned for Volume 2 include articles on the distribution of Recent foraminifera of the neotropical region; on oxygen isotope studies as applied to foraminifera; a critical review of planktonic foraminiferal zonation; on trophic relationships of foraminifera; on comparative distributions of living and dead foraminifera; and a review of the life and work of Joseph A. Cushman.

R. H. HEDLEY and C. G. ADAMS

British Museum (Natural History)
Cromwell Road
London SW7 5BD

Notice to prospective contributors

FORAMINIFERA will include papers on all aspects of foraminiferal research, both Recent and fossil, and the Editors will welcome suggestions for future review articles. Throughout this series the following terminology will normally be employed: foraminifer (singular), foraminifera (plural), foraminiferal (adjectival) and Foraminiferida (the ordinal name, previously known as Foraminifera).

Contents of Volume I

Contributors v

Preface vii

Recent advances in the classification of the foraminiferida
ALFRED R. LOEBLICH AND HELEN TAPPAN 1

Biometry of the foraminiferal shell
G. H. SCOTT 55

Field and laboratory techniques for the study of living foraminifera
ZACH M. ARNOLD 153

Towards understanding the niche of the foraminifera
JOHN J. LEE 207

A. d'Orbigny—a biographical sketch
Y. LE CALVEZ 261

Systematic Index 265

Recent Advances in the Classification of the Foraminiferida

ALFRED R. LOEBLICH JR.

Department of Geology, University of California, Los Angeles, California, 90024, U.S.A.
Chevron Oil Field Research Company, La Habra, California, 90631, U.S.A.

and

HELEN TAPPAN

Department of Geology, University of California, Los Angeles, California, 90024, U.S.A.

I.	The Nature of Classification.	1
II.	Changing Bases and Degree of Foraminiferal Classification	2
III.	Modifications and Taxonomic Revisions	6
	A. Suborder Allogromiina	6
	B. Suborder Textulariina	7
	C. Suborder Fusulinina	15
	D. Suborder Miliolina	17
	E. Suborder Rotaliina	19
IV.	Phylogeny at the Family Level	39
V.	Summary of Revised Classification	42
VI.	Acknowledgements	50
VII.	References	50

I. The Nature of Classification

"In classifying animals," states Webster's Unabridged Dictionary, "the attempt has usually been made to take into account all characters as far as practicable, and in accord with the doctrine of evolution, to show their natural relationships and lines of descent from common ancestors."

As we stated previously (Loeblich and Tappan, 1964a, p. C153), a classification of foraminifera ideally should be based on complete data concerning the morphology of test and cytoplasm, details of the reproductive process, ontogenetic changes, life habits and habitat, and geologic occurrence, in order to determine such relationships and lineages. As this ideal level of factual data can never be obtained, any generalizations as to natural relationships must be based on partial information, which is limited in varying degree.

The two main philosophical approaches to foraminiferal classification currently either involve a primary consideration and ordering of phenetic similarities, or involve a classification based on a previously constructed phylogenetic scheme. These philosophies would appear to be diametrically opposed, yet on closer examination their differences seem more semantic than factual. As noted by Hull (1964), phylogeny is not a fact to be discovered, but an abstraction that "is inferred almost exclusively from morphological, genetical, paleontological, and other types of evidence and is not observed directly." The evidence available for determination of phylogenies also may be incomplete and indirect, but in any event, it largely consists of phenetic similarity. In inferring phylogenetic history, the assumptions made as to the relative importance of various morphologic characteristics differ little from those made by zoologists in producing taxonomic schemes and classifications from a single time plane, and although a paleontologist has geologic data which add another dimension, he does not always impart to such data a major importance in classification. In fact, as stated by Raup and Stanley (1971, p. 130), a biologist or paleontologist rarely has "a truly accurate knowledge of phylogeny; classification is thus an expression of knowledge at a given time."

II. Changing Bases and Degree of Foraminiferal Classification

Since the description two centuries ago of *Ammonia* Brünnich, 1772, as the first generic name to be proposed specifically for a foraminifer, knowledge of this group has increased at an exponential rate. By the time the first 100 or so genera had been named, an attempt was made to arrange these in a systematic classification (de Blainville, 1825), although this classification into 10 families, based on test shape, included a mixture of foraminifera and cephalopods. After the true nature of the protozoans was demonstrated, foraminifera were regarded as simple and primitive organisms, and extremely wide limits were allowed in taxa of all levels. The classifications utilized by Brady (1884) and Cushman (1925) still had only 10 families, and a small number of subfamilies. During the following 50 years, taxonomic studies expanded rapidly, and in 1948 Cushman allocated the 770 recognized genera to some 50 families. Rauzer-Chernousova and Fursenko (1959) recognized 72 families and 91 subfamilies, and the Treatise on Invertebrate Paleontology (Loeblich and Tappan, 1964a) recognized 1192 genera (with some 1267 as synonyms). These were placed in some 94 families and 128 subfamilies.

The basis for classification, and the variations in these over the years has been summarized at length (Loeblich and Tappan, 1964a–c) hence needs no repetition. In their *Treatise*, "a complete revision of all families and genera of foraminifera could be made in the light of the recent new information concerning the ... important morphologic characters of test mineralogy, wall microstructure, lamellar and septal characters, internal partitions and apertural modification (toothplates) and the physiological data as to the living animal, life cycle and type of gametes" (Loeblich and Tappan, 1964c, p. 15). Many earlier genera had not been described with reference to wall and septal microstructure and internal and apertural modifications, and were redefined and emended extensively in the Treatise. Not all type species were available then for such restudy, hence certain older genera still await complete definition. In addition, in the 11 years since the Treatise went to press, new lines of research have been underway. These range from transmission and scanning electron microscopy to aid our understanding of the surface textures and porosity, wall crystal structure and lamination and apertural features, electron probe microanalysis to determine the presence and importance of trace elements, chemical analyses of the organic test wall or organic lining of others, a few life cycle and life history studies to delimit the ecologic controls of morphology, the relationships within polymorphic species and ontogenetic changes in others, to the more classical but still necessary studies with the light microscope of test exterior and, by means of thin section or dissections, the internal structures of the test.

In 1962, when the Treatise was submitted to the editor, a single publication had utilized electron microscopy in a study of a foraminifer, and the latter was not identified. A wealth of studies of foraminifera by means of electron microscopy have appeared in the decade since Hay *et al.* (1963) first made transmission electron micrographs of foraminiferal wall surfaces, and in fact scanning electron micrographs now are becoming routine for illustration and description of species.

Perhaps the emphasis on such ultrastructural studies, and to some extent an emphasis on reassessment of already described taxa, has been the cause of a slight decrease in the rate of description of new generic taxa. Thus, from an average annual production of 60 new genera proposed in the ten years preceding the Treatise an average of about 40 genera have been newly described annually during the last five years. Altogether, some 554 new generic taxa have been proposed in the 11 years to date since the Treatise, bringing the total number of generic names utilized for foraminifera to well over 3000. Of the 1267 of these regarded in the Treatise as synonyms, some subjective "synonyms" later were reinstated as valid on the basis of additional evidence or new criteria; a few that were tentatively recognized as valid in the Treatise have since been placed in synonymy (in fact, some of these have been variously allocated as synonyms to two or three different genera by different workers, this disagreement indicating the necessity of further and more detailed study).

Over the past dozen years, additional suprageneric categories proposed include some 40 new family names and 52 new subfamilies. The composition of many family group categories also has undergone repeated but generally minor revision. The new taxa at all levels, and the revised relationships postulated have involved most of the lines of research mentioned above that have concerned foraminifera during this period. As a result, a review of recent advances in foraminiferal classification also involves nearly every aspect of foraminiferal study. Perhaps of greatest importance have been the advances in interpretation of wall structures, resulting from studies of the surface and internal microstructure by means of transmission and scanning electron microscopy. Some previous observations made with light microscopy have been confirmed and elucidated with the new methods, and some conflicting interpretations have been resolved, for example, the mono- or bilamellar character of various hyaline foraminifera which had been the subject of endless arguments, and even differing interpretations of the same illustrated thin sections. When the monolamellar versus bilamellar character of the hyaline calcareous test, once regarded as of superfamilial importance (Reiss, 1958, 1963; Loeblich and Tappan, 1964a, 1964c) was demonstrated by electron microscopy to represent a matter of degree of thickness rather than one of presence or absence of separate laminae, the original basis for separation of the Discorbacea and Orbitoidacea (sensu Treatise), or of the Nonionacea and Anomalinacea (Loeblich and Tappan, 1964c) was invalidated, and their redefinition on other criteria requires reallocation of some families and genera. For example, redefinition of the Orbitoidacea might restrict this superfamily to the orbitoidal types with large embryonic gamonts and complex structure (Glaessner, 1963), while transferring other "Bilamellidea" to the vicinity of the Discorbacea.

Other problems involved in the classification of the foraminifera remain in contention, such as the degree of relationship of certain nominal genera or species with similar early ontogenetic stages, but highly distinctive adults; these questions will be resolved only by the accumulation of more complete and unequivocal life history data.

Some classification revision also has been undertaken on the basis of phylogenetic schemes, with varying degrees of success and acceptance. The continuing high rate of description of new taxa suggests that for many, if not most, foraminiferal categories, insufficient data are as yet available for determination of lineages. Even among the planktonic foraminifera, "well documented evolutionary lineages are still comparatively rare where ancestors grade into descendants; most species and subspecies appear in the Cenozoic without apparent ancestors" (Jenkins, 1973, p. 138). Postulated relationships based on occurrences in one geographic region may later be disproved elsewhere. *Globo-*

quadrina praedehiscens was regarded (Blow, 1969) as having given rise to *G. dehiscens* during the Miocene (*Globorotalia* (*T.*) *kugleri* Zone) but Jenkins (1973) reported that *G. dehiscens* is present in the Oligocene of New Zealand and Australia, only migrating to the Trinidad region by Early Miocene. An ancestor in the New Zealand area, as yet unnamed, was reported to occur in the *G. euapertura* zone. Additional examples were reported in which evolution in one area was followed by later migration of the species elsewhere, commonly with a delay of a faunal zone or more.

Far from being an isolated occurrence, such incomplete or inconsistent sequences of species might well be expected from the process of evolution, for "allopatric speciation in small peripheral populations automatically results in gaps in the fossil record " (Eldredge and Gould, 1972). Evolutionary theory suggests that a new species does not evolve in the main area of distribution of its ancestor but in an isolated population on the very margin of its occurrence. Thus, a complete local record of a gradual splitting of lineages would be highly improbable and would not be observable by tracing up a local rock sequence.

Construction of a phylogenetic scheme should be in agreement with, if not strongly influenced by, known geologic occurrence, but many proposed phylogenies seem to have ignored this, so that proposed "ancestral" taxa in fact are of considerably later appearance than their supposed "descendants".

Hofker (1971a) on the basis of certain apertural characteristics and suggested homologies, derived *Robertina* from *Conorboides*, and regarded these aragonitic-walled genera as ancestral to the granular calcitic Nonionidea, the porcelaneous Peneroplidae, the radial calcite Epistomariidae, and canaliculate Rotaliidae, Elphidiidae (Faujasininae and Elphidiinae), and Nummulitidae, in contrast to the *Treatise* classification in which wall microstructure and composition were regarded as indicative of monophyletic origin. In addition, *Robertina* is known only from Eocene and later strata, whereas the suggested descendants were all of geologically earlier appearance: Peneroplidae (Triassic?, Cretaceous— Holocene), Nonionidae, Epistomariidae, Cuvillierininae, Rotaliinae (all Cretaceous to Holocene), and the Nummulitidae (Paleocene to Holocene). Of these, the Peneroplidae, Elphidiinae (Eocene to Holocene), *Faujasina* (Pliocene), *Epistomaria* (Eocene) and *Alliatina* (Pliocene to Holocene) all were derived through *Cushmanella* (Holocene) by Hofker. Obviously, opinions differ as to the relative importance of various morphological criteria in ascertaining lineages, but such discrepancies in geologic age strongly counterindicate such phylogenies.

III. Modifications and Taxonomic Revisions

Some of the many suggested taxonomic and classificatory changes of the past decade are outlined below, under the Suborders Allogromiina, Textulariina, Fusulinina, Miliolina, and Rotaliina, as recognized in the Treatise. Little disagreement is encountered as to the major importance of shell wall composition, as similar groupings have been recognized for well over a century, but superfamilies of the Rotaliina, recognized in the Treatise on the basis of crystal form, optically radial or granular micro-structure, and septal lamellar character, have been variously questioned on the basis of new data from electron microscopy, or from the viewpoint of proposed phylogenetic schemes in which these features were regarded as repeatedly and independently derived in various lineages.

Haynes (1973, p. 220) regarded as unlikely the possibility that the various types of wall structure arose independently from tectinous ancestors, and considered the hyaline radial and granular types of wall to be derived from each other at various times, whereas he considered that the Treatise classification "required the independent origin of all the different wall structure groups from tectinous ancestors." We stated in the Treatise (1964a, p. C153) ". . . the same chamber arrangement and form may have developed in independent lineages by parallel evolution without indicating interrelationship of the similarly shaped shells," and elsewhere (Loeblich and Tappan, 1964b) in a diagram of the hypothetical origin of superfamilies, indicated the possible indepen-dent origin from tectinous ancestors of only the agglutinated wall, the porcelaneous type, and the calcareous microgranular one (Endo-thyracea); from the latter in turn evolved the Fusulinacea, Nodosariacea, Spirillinacea, hyaline radial and hyaline granular perforate calcareous foraminifera. More recent studies seem to uphold some of these latter derivations.

The number of genera proposed since the appearance of the Treatise totals well over half the total recognized in 1964, and the number of family group categories newly proposed approaches half the number then recognized, hence space limitations prevent complete discussion of all. Most new family categories and major revisions of earlier families are mentioned, but while a few examples of significant generic changes are included, a failure to mention others is due only to lack of space rather than to a value judgment at this point.

A. Suborder Allogromiina

These foraminifera with membranous or tectinous test have received little additional classificatory study, although some life history and cytological investigations are of interest.

Bermúdez and Rivero (1963) made a systematic revision of chitinous, microgranular and arenaceous taxa, all "chitinous" forms being included in the family Lagynidae, which included the subfamilies Lagyninae, Amphitrematinae, Myxothecinae, Allogromiinae and Rhynchogromiinae. Of these, the Amphitrematinae (as a family) was placed in the Order Gromiida in the Treatise, the Myxothecinae included in the Lagynidae, and the remaining two in the Allogromiidae.

Boderia turneri (studied by Hedley *et al.*, 1968) was found to have biflagellate gametes, as do other members yet reported in the family Lagynidae. Their study by electron microscopy, however, indicated that one of the unlike flagella was pantonematic, possessing two longitudinal rows of cilia. Such complex flagella are known in the euglenids, chrysomonads, and dinoflagellates, whereas green algae generally have identical smooth flagella. This fine structure may indicate phylogenetic relationships at the level of major categories.

Shepheardella taeniformis (family Allogromiidae) was restudied in culture (Douglas, 1964), and based on differences in the reticulopodia, Douglas reinstated *Tinogullmia* Nyholm as distinct from *Shepheardella*. Ultrastructural studies (Hedley *et al.*, 1967) described its cytoplasm and nucleus in detail, and indicated a spongy, non-oriented fibrous structure for the tectinous wall of *Shepheardella*.

B. Suborder Textulariina

This suborder includes all basically agglutinated foraminifera, although both the organic-walled Allogromiina and the porcelaneous Miliolina include representatives with occasional agglutinated foreign material in the wall. Major advances in recent years have been the increased understanding of the agglutinating cement, the nature and occurrence of an organic lining and variations in the nature of the foreign material utilized as well as the microstructure of the resulting wall. Some of this new information has been used in redefinition of genera and families, proposing new synonymies and for reinstatement as valid some genera formerly regarded as synonyms. Certain proposed changes in classification were in conflict with the Zoological Code, as will be noted, and must be modified.

The nature of the organic cement in the agglutinated wall was examined by Hedley (1963), and Hedley *et al.* (1967). According to Hedley (1963), an inner "chitinous" lining is not common, and in some recent species preserved cytoplasm adhering to the wall has been mistakenly reported as such (e.g., *Protobotellina cylindrica*). In other species the inner wall is covered by a thin cement layer (*Astrorhiza limnicola, Pilulina jeffreysi*). The cement is not chitin, but consists of a relatively stable acid or sulfated mucosubstance or glycoprotein (a combination of protein and carbohydrate), and may have organically bound iron, and probably calcium. Both the organic (amino acid) content and inorganic compounds (SiO_2, CaO, Fe_2O_3, Al_2O_3, MnO) were determined for *Pilulina jeffreysi, Protobotellina cylindrica, Hyperammina nodosa,* and *Pelosphaera cornuta.* The high amount of alumina found may have been erroneously "identified" as iron in earlier foraminiferal descriptions. Color also is not a reliable guide to iron content, as all examined species reacted positively to the test for iron, whether brown or white in color; the iron is indicated to be organically bound to the cement. Removal of all iron or calcium from the cement causes it to disintegrate, as the inorganic components appear vital to the stability and cohesive quality of the cement. The cytoplasm also showed a high concentration of iron in the peripheral areas in *Saccammina, Dendronina, Hyperammina* and *Protobotellina.* Shells that were hard and firm were mineralized with calcium, iron or a mixture of compounds during or after secretion of the organic cement; tests that remained soft and flexible lack such mineralization. A further modification was demonstrated in *Halyphysema* by Hedley *et al.* (1967). The spicules of the test wall are cemented by an acid mucopolysaccharide, but in addition, the inner surface of the basal disc and the lower quarter of the main body chamber is lined with a flexible, extracellular fibrous organic sheath of collagen. The fibers are 1 to 2 μm in diameter, and are composed of cross-striated filaments 0.2 μm in diameter. The collagen nature of these filaments was determined from the observations of its fine-structure in electron microscopy.

Hofker (1972) emphasized the importance of wall complexity in even the primitive agglutinated foraminifera, separating those with only one layer of sand grains in the wall (*Reophax, Armorella, Psammosphaera, Rhabdammina*) from those with many layers of grains (*Hyperammina, Hormosina, Astrammina, Loeblichopsis*), both on the level of genera and subfamilies.

The agglutinated foraminifera were divided in the Treatise into the superfamily Ammodiscacea (84 genera) consisting of a single chamber or proloculus and undivided second chamber, and the Lituolacea (209 genera) which are the polythalamous agglutinated forms. New family-group names recently proposed for the Ammodiscacea include the families Dendrophrynidae Vyalov, 1966; Flagrinidae Vyalov, 1966; Hippocrepinidae (ex subfamily of Rhumbler, 1895) Vyalov, 1968; Oxinoxisidae Vyalov, 1968; Silicamminidae Vyalov, 1968; Silicotubidae Vyalov, 1968; and Usbekistaniidae Vyalov, 1968, and the subfamilies Ammovolummininae Chernykh, 1967; Crithionininae Hofker, 1972; Marsipellinae Hofker, 1972; Tholosininae Bermúdez and Rivero, 1963; Thurammininae A.D. Miklukho-Maklay, 1963; and Usloniinae Miklukho-Maklay, 1963. In addition, the subfamily Pseudoreophacinae Suleymanov, 1963 was based on the genus *Pseudoreophax* Suleymanov, 1963 (non *Pseudoreophax* Geroch, 1961, and = *Adelungia* Suleymanov, in Arapova and Suleymanov, 1966).

Recently proposed family group names referable to the superfamily Lituolacea of the Treatise include the families Asanospiridae Vyalov, 1968; Aschemocellidae Vyalov, 1966; Coskinolinidae Moullade, 1965; Cystamminellidae Vyalov, 1968; Nezzazatidae Hamaoui and Saint-Marc, 1970; Textulariellidae Grönhagen and Luterbacher, 1966; and subfamilies Chrysalidininae Neagu, 1968; Coskinolininae Cimerman, 1969 (nom. transl. ex family of Moullade, 1965); Coxitinae Hamaoui and Saint-Marc, 1970; Dictyoconinae Moullade, 1965; Dorothiinae Balakhmatova, 1972; Hemicyclammininae Banner, 1966; Kurnubiinae Redmond, 1964; and Trochoporininae Soliman, 1972.

In their major and inclusive revision, Bermúdez and Rivero (1963) recognized two superfamilies and 19 families of chitinous, microgranular and arenaceous foraminifera. The Astrorhizidea contained the Astrorhizidae, Rhizamminidae, Saccamminidae, Hyperamminidae, and Tolypamminidae. The Lituolidea included the Reophacidae, Lituolidae, Textulariidae, Verneuilinidae, Orbitolinidae, Trochamminidae, Loftusiidae, Placopsilinidae, and Rzehakinidae (as in the Treatise approximately), but also included the Palaeotextulariidae and Tetrataxidae (of the Treatise calcareous microgranular Endothyracea), the Cymbaloporidae (in the Treatise placed with the hyaline Rotaliina), the Carterinidae (representing a distinct superfamily in the Treatise, on the basis of the secreted calcareous spicular wall), and the Neusinidae (which we had shown to represent the Order Xenophyophorida instead of the Foraminiferida, *Neusina* being a junior synonym of *Stannophyllum* Haeckel, 1899). Subfamily categories differed slightly, and a few were preoccupied, e.g., the Reophacidae and Reophacinae were preoccupied by Hormosinidae and Hormosininae, the Verneuilinidae by the Ataxophragmiidae, and the new subfamily Tholosininae by the Hemisphaerammininae.

In her manual of agglutinated foraminifera, Neumann (1967) utilized eight families: the Saccamminidae, Astrorhizidae, Hyperamminidae, Ammodiscidae, Reophacidae, Lituolidae, Textulariidae, and Ataxophragmiidae. Ninety selected genera were described and many illustrated by excellent photomicrographs of thin sections.

Vyalov (1966, 1968) described many new genera of large, straight to dichotomously branching, to enrolled tubular siliceous and agglutinated foraminifera, of which some were attached. Of the family group categories based on these forms, the monotypic prostrate attached Dendrophrynidae might better be considered a subfamily of the Astrorhizidae, paralleling the erect-growing attached Dendrophryinae, and might well also include *Sagenina* (from the Hemisphaerammininae in the Treatise), as well as *Flagrina* Vyalov, 1966 and *Aciculella* Vyalov, 1968 (=*Aciculina* Vyalov, 1966, non Adams, 1853) which both had been included by Vyalov in the Flagrinidae.

In the Silicamminidae, Vyalov isolated a few Cenozoic genera of simple form, *Silicammina* Vyalov, 1966, *Hirudina* Vyalov, 1966 and *Saccamminoides* Geroch, all from Eocene to Miocene age. The Silicotubidae Vyalov similarly included tubular siliceous tests, referred to *Bogdanowiczia* Pishvanova and Vyalov, 1967 and *Silicotuba* Vyalov, 1966, which strongly resembles various members of the Rhizammininae.

Miklukho-Maklay (1963) subdivided the Psammosphaeridae into the Psammosphaerinae, Stegnammininae and Usloniinae (as Uslonianinae). In the Treatise *Uslonia* was included in the Parathuramminidae (Parathuramminacea), and of the other members of the Usloniinae, *Raibosammina* and *Thekammina* had been regarded as synonyms of *Stegnammina* in the Treatise, and that genus and *Ceratammina* had both been included in the Psammosphaerinae, whereas *Shidelerella* we regard as a synonym of *Ordovicina* of the Saccammininae. In view of the disjunct geologic distribution of this group, it is highly possible that these morphologically simple forms are not closely related, and that the early Paleozoic monothalamous agglutinated genera could be better separated as the Family Stegnamminidae Moreman, 1930. The type genera and many others of the Saccamminidae and its subfamilies are based on Recent or at least Post-Paleozoic species. Many genera and the included species will need restudy for proper allocation if such separation is recognized. One such restudy of topotype material (by Conkin *et al.*, 1970) led to the recognition of *Weikkoella* (formerly of the Saccamminidae) as steinkerns, or internal molds of the primitive charophyte *Moellerina greenei*, and removal of the genus from the foraminifers. Miklukho-Maklay (1963) subdivided the Saccamminidae into two subfamilies, the Saccammininae with single aperture and rounded test, and the Thurammininae with many apertures, commonly produced on projections, or with irregular test. The latter was erroneously credited to Eimer and Fickert. Eimer and Fickert (1899, p. 600) had cited Thuramminidae Brady in the synonymy of their new Kyphamminidae, but in fact Brady (1884) had placed *Thurammina* in the Trochamminidae! Miklukho-Maklay included in both subfamilies a variety of agglutinated and microgranular calcareous genera (e.g., *Parathurammina, Saccamminopsis, Uralinella, Irregularina*) which we regard as more closely allied to the Fusulinina.

In his revision of simple agglutinated genera, Hofker (1972) recognized the subfamilies Astrorhizinae and Psammosphaerinae and reinstated the subfamilies Rhabdammininae, Reophacinae and Hyperammininae (presumably of the Family Astrorhizidae, which has priority among those included, although he did not specifically so state). However, *Botellina* was stated to merely contain agglutinated spicules that project into the chamber cavity rather than having a truly labyrinthic wall, and was thus considered congeneric with *Hyperammina*. Hofker regarded *Botellina* Carpenter, Jeffreys and Thomson, 1870, as a synonym of *Hyperammina* Brady, 1878, whereas in fact the reverse is true in view of priority. In addition, Hofker included in the Hyperammininae the additional genera *Saccammina*, *Ammodiscus* and *Hormosina*, so that the subfamily name would have been preoccupied by the Ammodiscidae Reuss, 1862, and also by the Saccammininae Brady, 1884, Hormosinidae Haeckel, 1894 and Botellininae Chapman and Parr, 1936. Two additional new subfamilies were proposed, the Crithionininae and Marsipellinae. *Marsipella* was redescribed as consisting of a proloculus and additional chambers somewhat resembling *Aschemonella*, rather than being simply fusiform or tubular; as so defined it would fit well within the Aschemonellinae. As defined by Hofker, the Crithionininae included *Hemisphaerammina*, *Pseudowebbinella*, *Sorosphaera*, *Hippocrepinella*, *Pilulina*, *Bathysiphon*, *Technitella* and *Chitinosaccus*. A subfamily with this composition would have the suggested name preoccupied by the Pilulininae Brady, 1884, and also by the Bathysiphoninae Avnimelech, 1952 and Hemisphaerammininae Loeblich and Tappan, 1961. Hofker reinstated some genera on wall features (*Armorella*, *Hippocrepinella*), newly proposed others (*Loeblichopsis*), and reduced *Saccorhiza* and *Tolypammina* to subgenera of *Hyperammina*.

The Subfamily Tholosininae of Bermúdez and Rivero included parts of the subfamilies Hemisphaerammininae, Diffusilininae, Psammosphaerinae, and family Lagynidae of the Treatise. We would separate those with complex wall as the Diffusilininae, and the Tholosininae also is preoccupied by the Hemisphaerammininae Loeblich and Tappan, 1961, as both subfamily definitions included the nominal genus of the other. An additional subfamily, the Ammovolummininae Chernyk, 1967, of the Ammodiscidae was proposed for two new straight and arcuate, elongate conical genera of the Silurian, *Ammovolummina* and *Serpenulina*. The Oxinoxisidae Vyalov, 1968, of Mississippian age, perhaps should be placed closer to the Hormosinidae, where its single genus had originally been placed, rather than within the Lituolidae (Placopsilininae) as in the Treatise.

Where Hofker had included single-chambered, tubular enrolled, and rectilinear polythalamous taxa in a single subfamily, on the basis of similar wall construction, Vyalov further subdivided many of the family groups. In addition to the Dendrophryidae, Flagrinidae, Silicamminidae and Silicotubidae, mentioned above, he proposed a new family Hippocrepinidae Vyalov, 1968 (ex subfamily of Rhumbler, 1895) for the genera *Hippocrepina* and *Giraliarella*, the new monogeneric Cretaceous Aschemocellidae Vyalov, 1966, the Usbekistaniidae Vyalov, 1968, for those Jurassic to Recent irregularly coiled "Ammodiscinae", *Usbekistania*, *Ammovertellina*, and *Arenoturrispirillina*, and the monogeneric Asanospiridae Vyalov, 1968, for the supposedly siliceous *Asanospira*, which we regarded as a synonym of *Haplophragmoides*. The Cretaceous genera *Cystamminella* and *Martyschiella*, both of Myatlyuk, 1966, and originally described within the Trochamminidae, were similarly placed in a new family Cystamminellidae Vyalov, 1968, based on the siliceous wall. In view of the evidence that many reportedly siliceous foraminifera in fact consist largely of organic "cement" in which foreign matter is embedded, and do not secrete silica, separate family status of the Usbekistaniidae, Asanospiridae, and Cystamminellidae appears of doubtful validity.

A more useful method of separating the agglutinated foraminifera concerns the presence or absence of wall perforations, long known, but never utilized systematically for classification or sufficiently in taxonomic description. The subfamily Trochoporininae Soliman, 1972, was proposed for the perforate-walled Trochamminidae. However, the type species of *Ammoglobigerina* Eimer and Fickert, 1899, was also included in the new perforate genus *Trochoporina*, as *T. globigeriniformis* (Parker and Jones); this distinct wall character provides a valid basis for reinstating *Ammoglobigerina* (with *Trochoporina* as a synonym) as distinct from *Trochammina*, whose type species was determined to be non-perforate. As yet, too few genera have been studied in section to properly use this character for family classification although it can be utilized wherever determined.

Balakhmatova (1972) revised the Ataxophragmiidae, recognizing three subfamilies, the Verneuilininae, Ataxophragmiinae and a new subfamily Dorothiinae. As circumscribed, the Verneuilininae include those genera placed therein in the Treatise, with the addition of other genera from the Ataxophragmiinae and Valvulininae (including *Valvulina*). Including *Valvulina* in this group would require the use of the Subfamily Valvulininae Berthelin 1880, instead of the Verneuilininae Cushman, 1911. Other suggested taxonomic modifications within this group include the placement of *Barbourinella* as a synonym of *Tritaxia*; *Paragaudryina* as a synonym of *Gaudryina* rather than of *Spiroplectinata*; *Gaudryinella* as a synonym of *Spiroplectinata*; recognition of *Pseudogaudryina* and *Siphogaudryina* as separate genera, rather than as synonyms of *Gaudryina*, and *Clavulinoides* as distinct rather than a synonym of *Tritaxia*; *Valvoreussella* was also placed in synonymy of *Migros* rather than of *Gaudryina*.

Chrysalidina and *Minouxia* had been removed from the Valvulininae to a new subfamily Chrysalidininae Neagu, 1968, which also included the new Lower Cretaceous (Barremian) *Andersenia* Neagu, 1968. This subfamily was regarded as related to the Valvulininae, but *Andersenia* is geologically earlier than any other Valvulininae, although it appears relatively advanced to have been ancestral to that group. Because multiple apertures appear to have arisen independently in many groups of the agglutinated foraminifera, we regard these genera as better placed within the Valvulininae than in a separate subfamily.

The new subfamily Dorothiinae of Balakhmatova, 1972, included part of the Treatise Globotextulariinae (exclusive of its type genus), Valvulininae, and Ataxophragmiinae, as well as the more recent genera *Pseudomarssonella* and *Riyadhella* Redmond, 1965, and *Eomarssonella* Levina, 1968. *Verneuilinella* was recognized as not a synonym of *Gravellina*, *Schenckiella* as not a synonym of *Martinottiella*, and *Marssonella* as not synonymous with *Dorothia*. The Ataxophragmiinae of Balakhamatova included genera variously assigned in the Treatise to the Ataxophragmiinae, Globotextulariinae and Valvulininae, as well as others more recently described, *Vialovella* Voloshina, 1972, *Arenobulimina* (*Columnella*) Voloshina, 1965 (a homonym of *Columnella* Levinsen, 1914), *A.* (*Novatrix*) Voloshina, 1965, *A.* (*Pasternakia*) and *A.* (*Harena*), both Voloshina, 1972, *Orbignyna* (*Ataxoorbignyna*) Voloshina, 1965, and *Ataxophragmium* (*Opertum*) Voloshina, 1965 (although including in the synonymy of this genus the prior *Pernerina* and *Ataxogyroidina*).

Hamaoui and Saint-Marc (1970) returned to the Lituolacea the family Barkerinidae (placed in the Miliolacea in the Treatise), stating that the imperforate calcareous microgranular wall might include varying amounts of agglutinated material. They revised the family as including *Barkerina*, as well as *Reticulinella* Cuvillier, Bonnefous, Hamaoui and Tixier, 1970, and *Praereticulinella* Deloffre and Hamaoui, 1970. Other genera placed therein previously were removed to a new family Nezzazatidae, with two subfamilies, the Coxitinae Hamaoui and Saint-Marc, 1970, containing *Coxites* and *Rabanitina*, and the Nezzazatinae, which included *Nezzazata*, and the new genera *Trochospira*, *Merlingina*, *Biconcava*, and *Biplanata*. Voloshina (1970) described an Albian *Barkerina* from the Russian Platform as porcelaneous, consisting of microgranular calcite which dissolved completely in weak acid, leaving no residue. Hamaoui (1973) restudied topotypes and determined true *Barkerina* to be monotypic, with a calcareous imperforate and agglutinated wall; the Barkerinidae thus are correctly placed in the Lituolacea, and the Russian species is referable to the porcelaneous *Ovalveolina* of the Miliolacea.

Agglutinated foraminifera with complex internal structure have been extensively revised (e.g., Redmond, 1964; Banner, 1966, 1970; Moullade, 1965; Grönhagen and Luterbacher, 1966; Hottinger, 1967; Cimerman, 1969).

Banner (1966, 1970) divided into three families the Lituolacea that have complex walls and internal pillars, but lack the "columella" of the Orbitolinidae and Pfenderinidae. The Lituolidae were defined as having a low initial spire, then commonly an uncoiled rectilinear portion; internal buttresses and partitions may be present, but no true labyrinthic or alveolar structure. The Ataxophragmiidae have a high trochospiral coil, and complex infolded walls may form internal buttresses and radial partitions to produce chamberlets. The third group he removed to a distinct family, the Spirocyclinidae, in which he included the six subfamilies, Choffatellinae, Loftusiinae, Hemicyclammininae, Cyclammininae, Spirocyclininae, and Mesoendothyrinae. Some genera were synonymized, others transferred to different subfamilies, and others reinstated as distinct genera. With the suggested composition of the family, however, the name would necessarily be the Loftusiidae Brady, 1884 (nom. transl. Lister in Lankester, 1903, ex Loftusiinae Brady). The major differences at the subfamily level were to split the Choffatellinae from the Cyclammininae, placing in the Choffatellinae *Choffatella, C. (Torinosuella)* which was reduced to subgeneric status and included *Pseudochoffatella* as a synonym, *Martiguesia*, and *Pseudocyclammina* (with *Paracyclammina* as a synonym); *Pseudocyclammina (Streptocyclammina)* and *Alveosepta*, both of Hottinger, 1967, also would be included here. The removal of *Paracyclammina* left the Loftusiinae monotypic. A new subfamily Hemicyclammininae included the new *Mayncella* Banner, 1966, and two genera from the Cyclammininae, *Hemicyclammina* and *Feurtillia* (and its synonym *Everticyclammina* Redmond, 1964); Hottinger, 1967, instead regarded *Mayncella* as a synonym of *Everticyclammina*. Probably *Rectocyclammina* Hottinger, 1967, and *Alveocyclammina* Hillebrandt, 1971, also would fall within this subfamily. The Cyclammininae was left with *Cyclammina, Alveolophragmium* and *A. (Reticulophragmium)*. Banner's Spirocyclininae was similar to that in the Treatise, with the omission of *Orbitammina*. The Mesoendothyrinae was reinstated, with a single included genus.

The Pavonitinidae of the Treatise included the subfamily Pavonitininae and the Pfenderininae. Cimerman (1969) showed *Pavonitina* not to have an initial triserial stage, and transferred it to the subfamily Tawitawiinae of the Textulariidae, the genus *Phyllopsammia* then becoming its junior synonym. The remainder of the Treatise subfamily Pavonitininae was redefined as the subfamily Coskinolininae Cimerman, 1969 (nom. transl. ex family Coskinolinidae Moullade, 1965), and the family name was replaced by the reinstated Pfenderinidae Smout and Sugden, 1962.

Moullade (1965) had revised the conical agglutinated genera of the Orbitolinidae, and erected the Coskinolinidae. The latter family included *Coskinolina* and *Lituonella* (from the Pavonitininae), *Kilianina* from the Pfenderininae, *Coskinolinoides* and the reinstated *Abrardia* from the Orbitolinidae, plus *Orbiqia* Mamgain and Jagannatha Rao, 1962; *Accordiella* might also be included here.

The Pfenderinidae (subfamily Pfenderininae) as restricted by Redmond, 1964, included *Pfenderina* and the new genera *Pfenderella, Sanderella*, and *Steinekella*; *Meyendorffina* was placed here by Redmond, but was in the Dictyoconinae of Moullade (1965). The subfamily Kurnubiinae Redmond, 1964, contained *Kurnubia*, *Pseudotextulariella* and the new genus *Praekurnubia*. Probably *Hensonia*, *Marieita* and *Pseudotriplasia* would also belong therein, but other genera were removed to the family Textulariellidae Grönhagen and Luterbacher, 1966, i.e., *Textulariella*, *Guppyella*, *Alveovalvulina* and *Alveovalvulinella*. These latter authors also considered *Cuneolinella* as a synonym of *Textulariella*, but we regard it as distinct. *Bronnimannina* Singh and Kalia, 1969, also belongs here. Because of the early trochospiral development, the Textulariellidae was believed close to the Ataxophragmiidae.

Moullade's revision (1965) of the Orbitolinidae recognized two subfamilies, the new Dictyoconinae, with *Dictyoconus*, *D.* (*Paleodictyoconus*) Moullade, 1965, *Meyendorffina*, *M.* (*Paracoskinolina*) Moullade, 1965, *Iraqia* and the reinstated *Orbitolinopsis* Henson, 1948 (previously included in the synonymy of *Orbitolina*). The Orbitolininae included *Orbitolina*, *O.* (*Praeorbitolina*) Schroeder, 1964; *Palorbitolina* Schroeder, 1963 (but as its type species, *Madreporites lenticularis* Blumenbach, 1805 is indicated in the Treatise as conspecific with *Orbulites lenticulata* Lamarck, 1816, the type species of *Orbitolina*, this genus is in fact a synonym of *Orbitolina*), *Neoiraqia* Danilova, 1963 and *Rectodictyoconus* Schroeder, 1964.

C. Suborder Fusulinina

The superfamilies Parathuramminacea and Endothyracea included many primitive Paleozoic taxa with microgranular calcareous wall and simple morphology. Redescription of some of these has resulted in their removal from the Foraminiferida and reassignment to other groups. For example, *Umbellina* was redescribed as a primitive charophyte (Miklukho-Maklay, 1961, Poyarkov, 1965, Mamet, 1970). The subfamily Umbellininae, now the family Umbellinaceae, includes *Umbellina*; *Elenia* and *Quasiumbella* of Poyarkov, 1965; *Pseudoumbella* and *Quasiumbelloides* of Berchenko, 1971, and *Biumbella*, *Lagenumbella* and *Protoumbella* of Mamet, 1970. Not all charophyte specialists would agree to this reassignment, but the occasional presence of an operculum-like structure on representatives of three of the genera appears to eliminate any possibility of foraminiferal affinity. The remaining genera of the Umbellininae of the Treatise (*Eolagena*, *Eovolutina*) may be transferred to the Tuberitininae.

Similarly, from the attached Ptychocladiidae, Mamet and Rudloff (1972) transferred the Stacheiinae (with included genera *Stacheia*, *Stacheioides*, *Fourstonella*, and *Aoujgalia*) to the family Ungdarellaceae of the Rhodophyta. Probably *Palaeonubecularia* and *Tscherdyncevella* similarly should be placed with this group of red algae.

New family groups proposed for the Endothyracea include the Robuloidinae Reiss, 1963, Endothyranopsidae Reytlinger, 1966 (elevated from the subfamily of Reytlinger, 1958); the Neotuberitininae, Lituotu-bellinae and Endothyrininae of Miklukho-Maklay, 1963; the Loeblichi-idae and Quasiendothyridae, elevated from subfamily status by Rozovskaya, 1963; and the Eostaffellidae Mamet in Mamet *et al.*, 1970. Some of these were erected for only two or three genera and do not appear sufficiently distinct to warrant such rank, whereas others were pre-occupied by other family group names that have priority.

Reiss (1963) defined the Robuloidinae as a subfamily of the Nodo-sariidae, including the genera *Robuloides* and *Pararobuloides* (regarded as in the Loeblichiinae of the Treatise) and *Protopeneroplis* (tentatively placed in the Involutinidae). The first two are very thick walled Permian taxa which appear close to the Archaediscidae, and probably would be best placed as a subfamily therein; they do not appear related to the Nodosariidae as placed by Reiss. Similarly, the Jurassic *Protopeneroplis* seems better retained in the Involutinidae.

The Loeblichiidae (as Loeblichinidae) and Quasiendothyridae were both indicated as having family status by Rozovskaya in Poyarkov, 1963; the Quasiendothyridae (ex subfamily of Reytlinger, 1961) was then utilized rather than Loeblichiidae (ex subfamily of Cummings, 1955), although the latter has priority. Included in the family were *Quasiendo-thyra*, *Q.* (*Eoquasiendothyra*), *Q.* (*Eoendothyra*), *Q.* (*Klubovella*), *Plano-endothyra*, *Loeblichia*, *L.* (*Urbanella*) Malakhova in Poyarkov, 1963, and *Dainella*. The Eostaffellidae Mamet in Mamet *et al.*, 1970 included *Eostaffella* (with *Paramillerella* as a synonym), *Eostaffellina*, elevated from *Eostaffella* (*Eostaffellina*) Reytlinger, 1963, *Loeblichia* and *Mediocris*, hence this family name is preoccupied similarly by the Loeblichiidae.

A major publication for the Fusulinacea (Kahler and Kahler, 1966a–c, 1967a) indexed all levels of fusulinacean taxa and included complete synonymies and occurrence records. The classification was more detailed than that of the Treatise 4 families and 5 subfamilies. Kahler and Kahler (1967a, b) recognized Staffellidae, with subfamily Staffel-linae and an unnamed "form-group containing *Pseudoendothyra* (the family Pseudoendothyridae Mamet, 1970, was later defined for this category); the Verbeekinidae, including the subfamilies Verbeekininae, Misellininae, Pseudodoliolininae Leven, 1963, Kahlerininae Levin, 1963, and Cheniinae Kahler and Kahler, 1966; the families Ozawain-ellidae; Schubertellidae with subfamilies Schubertellinae and Boulto-niinae; the Fusulinidae, with subfamilies Fusulinellinae, Fusulininae, Eofusulininae, and Wedekindellininae Kahler and Kahler, 1966; two subfamilies Polydiexodininae and Chusenellinae Kahler and Kahler, 1966, were recognized as of uncertain family assignment; the family Schwagerinidae, including subfamilies Schwagerininae and Pseudo-schwagerininae Chzhan, 1963; and family Neoschwagerinidae, with Neoschwagerininae and Sumatrininae.

The generally similar classification of Rozovskaya (1969) differed in some respects. An elaborate phylogenetic scheme was constructed that included both genera and family-groups. Rozovskaya included within the Fusulinacea the family Quasiendothyridae (including *Loeblichia*, and as noted above preoccupied by the Loeblichiidae); the Ozawainellidae, with subfamilies Ozawainellinae and Pseudostaffellinae; the Schubertellidae, with Schubertellinae and Boultoniinae; the Fusulinidae, with the same subfamilies as above except that the taxa of the Wedekindellininae were retained in the Fusulinellinae; the Schwagerinidae consists of the Schwagerininae, Polydiexodininae and Pseudofusulininae (which included the Kahler's Chusenellinae); the Staffellidae (unsubdivided but including the Kahler's Cheniinae and Kahlerininae, the Pseudoendothyridae of Mamet and the Nankinellinae Miklukho-Maklay, 1963); the Verbeekinidae, with the subfamilies Verbeekininae and Misellininae (including the Kahler's Pseudodoliolininae); the Neoschwagerinidae, similarly including the Neoschwagerininae and Sumatrininae, and in addition the Thailandininae Toriyama and Kanmera, 1968 (including the new genera *Thailandina* and *Neothailandina*, both of Toriyama and Kanmera, 1968).

D. Suborder Miliolina

Basic information as to the composition and microstructure of the porcelaneous wall has come from electron microscopy in recent years. Transmission electron microscopy of carbon replicas (Hay *et al.*, 1963) showed the wall of *Quinqueloculina seminulum* to consist of fine, randomly oriented rod-like crystals, $\frac{1}{2}$ to 2 μm in length and $\frac{1}{4}$ μm diameter, with a surface layer of organic matter, or with a surface layer of similarly oriented tabular crystals about $1\frac{1}{2}$ to 2 μm in diameter and $\frac{1}{2}$ μm in thickness arranged in a tile-roof pattern. These crystals were regarded as tabular rhombohedra ($10\bar{1}1$). Other Miliolacea had similar ultrastructure, although not all had the oriented surface layer. The perforate protoconch of the peneroplids, alveolinids and keramosphaerids was suggested as a possible basis for their separation from other imperforate porcelaneous taxa. Since then, the Miliolina have been emended (Brönnimann and Zaninetti, 1971) to include all porcelaneous species, both perforate and imperforate, in particular to include a new Triassic taxon with perforate walls for which the new family Milioliporidae was proposed.

Microstructure like that of *Quinqueloculina* was described (Lynts and Pfister, 1967) in *Miliolinella subrotunda* and in *Archaias*, which also had pseudopores in the pavement layer of tabular crystals that ended blindly in the randomly oriented layer. The three-dimensionally random crystal arrangement and orientation in the wall is unique in biologic systems in lacking a preferred orientation (Towe and Cifelli, 1967), and was believed to result from an incoherent colloidal active organic ground mass present in the crystallizing solution. The crystals fill spaces in the passive organic membranes that give shape to the test. The crystallographic randomness is due to the inability of the active organic matrices to form a uniform spatial macromolecular arrangement, except in a limited area. Colloidal organic matter trapped in the elongate crystals might stabilize the calcite structure, as a relatively large amount of magnesium is incorporated in the solid solution. The milky opacity (porcelaneous appearance) of the wall results from the completely random nature of crystal units and hence the refraction of light in all directions.

Scanning electron microscopy of the surface indicated a still greater variety of surface patterns in species of *Quinqueloculina* and *Triloculina*: the tile-roof pattern of surface rhombohedral crystals covering the random array of calcite needles; a mosaic pattern of equidimensional but somewhat irregular quadratic crystals of varied size; a brick pattern consisting of a subparallel arrangement of rhombohedra, and a parquet pattern with densely arranged elongate crystals; and a cobble pattern, with crystals normal to the surface overlying a three dimensional array of calcite needles. As this variety occurred in species of only one or two genera, and in some instances differed in parts of a single test, the superficial patterns were regarded as environmentally rather than genetically controlled, hence not of value for classification (Haake, 1971).

A reclassification of the Miliolina (Anglada and Randrianasolo, 1971) combined the Fischerinidae and Nubeculariidae in the latter family, while recognizing the same subfamilies as in the Treatise, except that the Calcivertellinae was included within the Cyclogyrinae and the Spiroloculininae transferred to the Miliolidae. The Miliolidae, Soritidae, Alveolinidae, Barkerinidae, and Squamulinidae remained largely as they were.

The Family Hemigordiopsidae Nikitina, 1969, included *Hemigordius* and *Hemigordiopsis*, and provisionally *Meandrospira*. It was defined as including calcareous imperforate taxa with initial globular proloculus followed by a streptospirally wound tubular chamber that later became planispiral. The tubular chamber may be subdivided into pseudochambers by rudimentary septa, these rudimentary septa and the streptospiral coiling eliminating this group from the Fischerinidae, Nubeculariidae and Miliolidae. Although *Agathammina* was suggested as perhaps ancestral to the Hemigordiopsidae, possibly both that genus and *Flectospira* should be transferred therein.

The Family Milioliporidae Brönnimann and Zaninetti, 1971, was erected for their new Triassic genus *Miliolipora*, whose porcelaneous wall is pierced by large subcircular pores at all ontogenetic stages, rather than only in the proloculus, as in *Peneroplis* and others.

The Barkerinidae, placed in the Miliolina in the Treatise was divided into Barkerinidae and Nezzazatidae by Hamaoui and Saint-Marc (1970) and replaced in the Lituolacea because of the presence of some agglutinated matter. As noted (p. 13 herein) under Lituolacea, the porcelaneous *Barkerina* described by Voloshina (1970) was reassigned to *Ovalveolina* by Hamaoui (1973). These species were only superficially similar, and as the facies in which they occur commonly is subject to recrystallization, the resultant poor preservation led to misidentification.

E. Suborder Rotaliina

The most extensive recent changes related to classification result from new information on the primary characters of wall structure and septal lamination of the hyaline calcareous foraminifera, i.e., septal lamination, radial or granular microstructure, and extent of perforation. In polarized light those hyaline perforate foraminifera whose extinction pattern results in a black cross with colored rings are said to have a perforate-radial microstructure, the c-axis of the calcite or aragonite crystals being oriented normal to the test surface. Other hyaline foraminifera show tiny flecks of color, suggesting a minutely granular structure, when viewed between crossed nicols; only calcite occurs in this perforate-granular type. Such wall structures were regarded as of superfamily rank in the Treatise, and the radial and granular groups believed to have originated independently from the Late Paleozoic Endothyracea (Loeblich and Tappan, 1964b). Seemingly transitional forms have now been found in the Permian-Lower Jurassic. However, the validity of a distinction between the radial and granular microstructure is still questioned, in part because of a dual usage of the terms (Towe and Cifelli, 1967). The distinction originally was made on the basis of optical phenomena in polarized light, indicating for the radial structure a preferred orientation of c-axes normal to the wall surface. The fibrous appearance of the radial wall was then considered as indicative of regularly aligned elongate crystals; in contrast, the flecks of light seen in the granular wall were regarded as due to irregularly shaped crystals of random arrangement. As the terms radial and granular also came to have a morphologic connotation, as seen in the light or transmission electron microscope, they became meaningless for classification, with both types of "wall structure" being reported in a single species (Bé and Ericson, 1963; Pessagno, 1964; Pessagno and Miyano, 1968; Hansen *et al.*, 1969). Some species had an inner "granular" wall layer, constructed of 0.5 μm grains, and an outer crust of prismatic crystals. Others (*Globorotalia, Pulleniatina*) had a test wall with columnar structure, but septa with granular texture although the latter also had an optically radial appearance in polarized light (Hansen *et al.*, 1969).

When the terms radial and granular are restricted to use in a crystallographic sense, with reference to the optical effect, a sharp distinction is nonetheless possible. Electron microscopy has shown that the fibrous structures are not single crystals in the sense used by X-ray crystallographers. All crystals of more than a few unit cells are subject to defects in structure, although at low resolution such defects are overshadowed by the statistically orderly molecular configuration. Apparently uniform crystals as seen in the polarizing microscope, are shown by electron microscopy to represent mosaics of smaller crystals. In the radial structure, crystal growth preferentially occurs along any axis of binary symmetry; the (0001) or basal pinacoid is the plane of growth fixed by epitaxy, and the c-axis is vertical to the organic matrix which serves as a template, and thus normal to the wall surface (Towe and Cifelli, 1967). Whether a crystal, a grain, or a mosaic of grains is involved, uniform extinction will occur under crossed nicols if the crystallographic axis is perpendicular to the stage. Thus, even a morphologically "granular" wall, such as the septa of the planktonic foraminifera mentioned above, would appear optically radial. The hypothesis of the (0001) orientation in radial foraminifers (Towe and Cifelli, 1967) later was corroborated for *Polymorphina* with the X-ray diffractometer (Hansen, 1968).

The organic matrix was regarded as consisting of a passive substance (polysaccharide-like) which provides the form of the chamber, and an active substance (protein), that provides a site for mineralization (Towe and Cifelli, 1967). The organic matrix consists of a three-dimensional spongy framework (Hansen, 1970b), and crystals develop until stopped or inhibited by the secretion of additional organic matrices. Crystals are interrupted at the resultant lamellar boundaries, nucleation being renewed on the other side of the boundary with the same crystal orientation.

Hansen (1970b) distinguished between the optically radiate-ultrastructurally radiate forms (*Nodosaria, Polymorphina, Bulimina*), and those with optically radiate-ultrastructurally nonradiate microstructure (*Ammonia*), in addition to the optically granular ones (*Melonis, Heterolepa*). In the optically radiate-ultrastructurally radiate species examined, organic membranes enveloped the single crystal units of the wall, and spongy organic material was also concentrated between the secondary lamellae.

On the basis of the morphological granularity, Hofker (1969) suggested that the granular calcareous wall was the result of included foreign matter such as coccoliths, and as such foreign matter was less available to the plankton, their walls were radially constructed. Benthic species, such as *Planulina*, were reported to have separate granular, calcareous fibrous radial, and crystalline radial wall layers (Hofker, 1970). However, as suggested above, the granular wall also has a preferred crystal orientation (Towe and Cifelli, 1967), but the c-axis is never parallel or perpendicular to the test surface. If crystal orientation were completely random, as it appears optically, the granular wall would be opaque like the porcelaneous one. Therefore, as the granular wall also must have a preferred orientation, nucleation of the calcite must be in a plane that does not allow the optical crystallographic axis to be fixed parallel or perpendicular to the test wall, that is, epitaxial growth on any (hkil) or hexagonal pyramid face. It is most likely to be on any of the (h0h̄1) rhombohedral faces characteristic of calcite. Towe and Cifelli (1967) further state that calcite probably grows on the hexagonal (101̄1) face (rhombohedron), hence the c-axis is oblique to the wall surface. With the X-ray diffractometer, Hansen (1968) showed the granular perforate *Melonis scaphum* to have the cleavage rhombohedron (101̄4) oriented parallel to the wall surface. One optical crystal unit in the wall of this species consists of a morphological single crystal, bounded by an organic membrane, just as in the radial type of wall. Pores were present both within the crystal unit and between them (Hansen, 1970b). In another granular genus, *Heterolepa*, crystal units are composite, and each tiny calcite plate is bounded by a delicate organic membrane, thinner than that surrounding the entire crystal unit. Instead of an optically and morphologically single crystal in a crystal unit as in *Melonis*, *Heterolepa* has crystal units formed of a group of plates with the same optical and morphological orientation. The piles of calcite plates have different orientation on opposite sides of the thick organic matrix, which may cause the indistinct extinction pattern (Hansen, 1970b).

Because the difference between radial and granular walls thus depends upon whether crystal growth is on the (0001), with vertical c-axis, or on the (101̄1), with oblique c-axis, a change from radial to granular microstructure could occur by a stereochemical modification in the active organic matrix, with epitaxial significance and could have occurred more than once. ". . . The only limitation that can be imposed with certainty is that radial and granular structures could not occur on the same chamber . . . [but] it is theoretically possible for transformation to occur during the ontogeny of an individual" (Towe and Cifelli, 1967, p. 755).

In many instances, generic allocation, and family placement has been based on the radial or granular nature of the wall structure, such placement commonly being corroborated by a variety of morphological criteria, apertural structures, etc. However, Hansen (1972) reported two species of *Turrilina* to be nearly identical in the light microscope and in the scanning electron microscope, but the older *T. brevispira* from the Lower Eocene was granular as seen in crossed nicols, whereas the middle Oligocene *T. alsatica* was radial. He suggested that the evolutionary change from a granular to radial structure in this genus occurred within the early Oligocene, the wall character being only of specific value, and not sufficiently stable for use in classification. However, he also noted that the possibility of convergent evolution could not be wholly eliminated. Haynes (1973, p. 196) similarly included both radial and granular species in *Elphidium*, and combined the Elphidiidae with the Nonionidae, regarding wall character as progressively changing, the more advanced forms of the lineages tending to be optically radial.

The septal lamellar character of the hyaline calcareous foraminifera also was regarded (in the Treatise) as of high level taxonomic significance, separating those with monolamellid septa (Nodosariacea, Buliminacea, Discorbacea, Cassidulinacea, Nonionacea) from the primarily bilamellar ones (Orbitoidacea, Globigerinacea, Anomalinacea) and the secondarily bilamellar ones (rotaliid septa) of the Rotaliacea. Limits of resolution led to conflicting opinions as to the true septal lamellar character of many, and in some instances, differing definitions or different interpretations of the same figures added to the confusion.

Hofker (1971b) stated that in the planktonic foraminifera, a primary organic membrane is formed first for the new chamber, and a thin (2 to 5 μm) hyaline calcite layer is deposited within the membrane. A second lamella of hyaline calcite is later secreted over the exterior and may continue to thicken. He thus considered all planktonic species to be primarily monolamellar. In contrast, Hansen and Reiss (1971, 1972), showed with electron microscopy that "monolamellar" and "bilamellar" taxa both had a median organic lining separating calcareous layers on either side, and that they differed largely in the thickness of the median organic layer and relative thickness of the inner and outer calcareous layers, the rotaliids differing in having a septal flap.

In the rotaliid *Ammonia*, the thin median layer consists of relatively large calcite crystals enveloped in organic matter, and arranged on each side of the organic median sheet. External to this is an outer lamella, whereas inside is an inner lining, both calcareous layers being constructed of stacked calcite platelets, the stacks arranged normal to the test surface. The platelets of each stack lie at an angle to the median organic layer and to the surface, the orientation of the platelets conforming to a crystal face of the median layer, those of adjacent stacks oriented about 110° apart to produce a herringbone pattern. The radial-walled *Cibicides* has paired calcite crystals in the median layer, whereas the granular-walled *Heterolepa* has euhedral crystals of heterogeneous size that are not paired. All have the stacked calcite platelets of the inner lining and outer lamella (Hansen and Reiss, 1971). The inner lining and outer lamella, and septal flap were interpreted for *Planulina* and *Heterolepa* as "pseudo-trilamellar" by Gonzàles-Donoso (1969).

In *Ammonia*, and other rotaliids, the inner lining occurs on lateral chamber walls and posterior wall, attached to and covering the preceding septal face (septal flap). Where the inner lining is not attached to the previous septal face, fissures or intraseptal passages are left. The inner lining also forms the plate-like "foraminal plate" (variously termed apertural lip or toothplate) around a partially resorbed foramen, and may extend into the preceding chamber to form the "umbilical cover-plate" (toothplate, or axial plate). The chamber wall, septal flap, foraminal plate and coverplate are formed by the inner lining as a continuous structure. The outer lamella covers the inner lining in the septal and lateral chamber walls including the lip, as well as the septal flap where it is not attached to the previous septal face; it also covers the inner lining of the foraminal plate and coverplate, and covers the outer lamellae of exposed earlier parts of the test. This produces a lamellar pattern in the wall as well as lamellar ornamentation (which is invariably of the "inflational" type, never discontinuous or "incised" as formerly believed from the light microscope). The median layer separates the inner lining and outer lamella where they are in direct contact, but does not separate successive outer lamellae or separate the inner lining and outer lamella of the previous chamber to which it is attached. The structure of *Rotalia* is similar to that of *Ammonia* in having an inner lining, median layer and outer lamella, but is more primitive, possessing only the septal flap, and lacking the foraminal plate and coverplate (Hansen and Reiss, 1971).

The median dark line in the septa of the bilamellar foraminifera was shown (Hansen and Reiss, 1972) to consist of an organic spongy layer, that occasionally included euhedral calcite crystals, separating the outer lamella and inner lining. By electron microscopy, a similar structure also was found in the "monolamellar foraminifera", but the median layer was thinner. In *Asterigerina*, and in the Robertinacea, the inner lining forms toothplates comparable to the septal flap and foraminal plate and coverplate of the rotaliids. Similarly, the Globigerinacea were shown to be bilamellar, the thin median layer being emphasized by the constriction of the pore tubules at this position, although the reliability of this character is disputed by Towe (1971). All planktonic species show both a primary lamination (multiple, thin layering) due to the calcite plates, and a secondary lamination due to the periodic growth.

Thus, the electron microscopy of the wall suggests that all perforate lamellar foraminifera are basically bilamellar, although those with thin median organic layer may appear to be monolamellar with the limited resolution of the light microscope (Hansen and Reiss, 1972). The Rotaliacea differ in the presence of the septal flap. As a result, certain reallocation of families and genera may be required within the redefined superfamilies.

1. SUPERFAMILY NODOSARIACEA

The Nodosariacea have generally been regarded as a relatively primitive group of the hyaline foraminifera, whose included genera have changed little since their origin in the Late Paleozoic. Studies of wall structure, lamination and perforation indicate a more gradual evolution from the Endothyracea during the Permian, Triassic and Early Jurassic (Gerke, 1957; Brotzen, 1963; Civrieux and Dessauvagie, 1965; Norling, 1966, 1968, 1972). On this basis, many new genera have been proposed for the older species, and others that had been synonymized have been reinstated.

Modern Nodosariids show two types of lamination, a primary or crystallographic lamination, due to the aggregates of flattened plates of calcite in the external wall and septa, and a secondary lamination in which the secreted layers result from recurrent growth (laminae added each instar), and which is not present in the septa. The primary lamination, which may be indistinct and observable only with polarized light, was suggested to be of the aggregation type (Civrieux and Dessauvagie, 1965), and not secretory calcite. Characteristic of Permian and Triassic species, it is rare in the Liassic ones, being present only in *Ichthyolaria* and *Mesodentalina*. The secondary lamination is not present in the Paleozoic Nodosinellidae (Endothyracea) but develops gradually from the non-lamellar wall by extension of the new chamber wall back over part of the preceding one, until finally each successive lamella completely envelops the entire test exterior. This superposition of secreted calcite occurs in three different ways in Liassic Nodosariidae, a secondarily nonlamellar type in which the new lamella is attached to only part of the preceding chamber and which is dominant in Liassic species (e.g., *Prodentalina*); an intermediate secondarily mesolamellar type, in which the new lamella covers one or more preceding chambers but does not completely envelop the test, as in *Mesodentalina* and some *Lenticulina*; and in the secondarily lamellar type the new chamber wall envelops and adheres to the entire test surface (in the Liassic present in some *Nodosaria*, *Dentalina* and some *Lenticulina*, but dominant thereafter) (Norling, 1968).

Within the Late Permian, 90 percent of the assemblage consists of the compressed Nodosinellidae, such as *Lunucammina* and *Pachyphloia*, with only primary lamination; the few symmetrical, uniserial, rectilinear, true Nodosariidae that are present similarly have only primary lamination, but may have intercalated limonitic or organic dark layers between lamellae. By the early Jurassic, secondary lamination is present in some Nodosariidae, and becomes progressively more important.

Electron microscopy has shown parts of the wall of the hyaline perforate foraminifera to be in fact imperforate, in particular the apertural periphery, sutures, septa and ornament. The entire test of the Paleozoic

Nodosinellidae was imperforate, as were early Nodosariidae, such as the Liassic *Geinitzinita* and *Astacolus*. Later nodosariids, such as *Nodosaria affinis* have imperforate ribs, with different extinction pattern in polarized light than that of the perforate interrib area (Hay *et al.*, 1963). The Jurassic *Ichthyolaria*, reinstated as distinct from *Frondicularia* in having only primary lamination (Brotzen, 1963; Civrieux and Dessauvagie, 1965; Loeblich and Tappan, 1964c; Norling, 1966) has ribs of imperforate granular microstructure which penetrate the fibrous radiate perforate inter-rib area, resulting in an alternation of granular and fibrous sections in the wall as seen in cross section (Norling, 1966). The degree of secondary lamination is correlated with the presence, size and frequency of pores in the wall. Those which are not secondarily lamellar are imperforate, or have very small pores (less than 1 μm diameter), or pores between 1 and 2 μm that are very sparsely placed. Secondarily lamellar forms have a high frequency of pores of 1 to 5 μm diameter, suggesting a pore function related to the formation of the lamellae. Because the pores of Jurassic Nodosariidae are similar in size to those of the Cretaceous, Tertiary or Holocene representatives, no evolution from protopores to deuteropores in the sense of Hofker is evident in this group (Norling, 1968).

Four types of wall microstructure occur in the nodosariids, of which the vesicular and granular types appear in the ornament regions and a thin inner wall layer, whereas the acicular and fibrous types appear in the inter-rib areas. The vesicular texture appears granular to magnifications of $\times 500$, but at greater magnification many hollow vesiculi of 1–10 μm size are discernible. Found in ribs and other surface ornament, this texture grades inward into the granular type. The granular texture consists of small (1–10 μm) closely packed polygonal grains or small globulites, commonly in a reticulate pattern; it occurs in the ornament of Lower Jurassic Nodosariidae and in an inner wall layer of some of these, in which it completely covers inner surfaces of wall and septa. This may be the first calcified layer outside the basal organic layer; as a similar inner granular layer occurs in some Paleozoic nodosaroid forms, and does not appear in any post-Liassic forms, it is regarded as primitive (Norling, 1968). The remaining two textural types, acicular and fibrous, characterize the exterior main part of the wall, the radial calcite consisting of fibrous or acicular crystals with the elongate *c*-axis normal to the test surface. When present, pores occur between the radially arranged crystals, hence are also radiating. The acicular structure is found in Liassic *Prodentalina* and *Astacolus*, whereas the fibrous type is most common in all other nodosariids. Although several genera have been restudied and subdivided on the basis of their wall characters, others require additional data. For example, Permian *Astacolus* have a compound test wall, others from Permian to Liassic may be non-lamellar and mesolamellar, whereas only completely lamellar forms are known from the Mid Jurassic to present (Norling, 1972).

The nature and systematic importance of the radiate aperture in the Nodosariidae has also been studied at length. The radiate aperture was suggested to be a reflection of the earlier septation of the chambers of the Paleozoic *Multiseptida* and *Colaniella* (Brotzen, 1963, p. 68); this is supported by the relation of the granular texture and penetrating nature of the rib ornament and circumapertural region of *Ichthyolaria* (Norling, 1968). In the Late Permian *Colaniella cylindrica*, the penetrating ornament forms ribs at the surface and also projects inward as septation.

Some early Nodosariidae have been reported as having radial apertures, and others as rounded, although the latter is commonly due to resorption. Four types of radiate aperture are reported in Jurassic species (Norling, 1968), those with radiating grooves or narrow, suture-like depressions leading to the opening (*Nodosaria, Prodentalina, Pseudonodosaria* and the Polymorphinidae); radiating slits which result in a stellate aperture (*Ichthyolaria*, some *Lenticulina*); apertural chamberlets with radiating slits from a central cap of shell material (some Jurassic *Prodentalina*, and Cretaceous and Cenozoic species), the chamberlet resorbed from early ontogenetic stages of older species, but persistent in younger ones; aperture with textural radiation, alternating fibrous and granular sectors bordering the opening, the granular sectors forming the ribs or costae (many *Nodosaria, Mesodentalina*), either projecting into the circular opening as spikes, separating the radiating (sometimes bifurcating) slits, or meeting to form an elevated ring around the aperture, and leaving only a rounded opening (Norling, 1968, 1972).

In the Treatise, this superfamily contained three families, the Nodosariidae, Polymorphinidae, and the Glandulinidae. The first of these, the Nodosariidae, included three subfamilies. The Nodosariinae, with straight and coiled genera, with radial or rounded aperture was the largest group with 37 genera; the Plectofrondiculariinae (3 genera) was initially biserial, and had a multiple aperture; whereas the Lingulininae included nine coiled or straight genera with slit-like aperture. Many other classifications have separated the straight and coiled forms in distinct families or subfamilies (e.g., those of d'Orbigny, Schultze, Reuss, Schwager, Rhumbler, Schubert, and Galloway cited in the Treatise). Putrya (1970) separated the enrolled forms as a "Family Lenticulinidae Sigal, 1952" [recte Family Vaginulinidae Reuss, 1860], containing four subfamilies. This family was regarded as having a distinct ancestry in the Late Paleozoic from that of the rectilinear Nodosariidae. The subfamily Lenticulininae [credited to Sigal, 1952, but recte Chapman, Parr and Collins, 1934] contained *Lenticulina, Robulus, Palmula, Neoflabellina*, and *Pravoslavlevia* Putrya, 1970. The new subfamily Astacolinae Putrya, 1970, contains *Astacolus, Saracenaria*, and *Dainitella* Putrya, 1972; the

Vaginulinopsinae Putrya, 1970 [recte Marginulininae Wedekind, 1937, nom. transl. Nørvang, 1957, ex family], included *Vaginulinopsis, Marginulinopsis*, and *Marginulina*; and the Planulariinae Putrya, 1970 [*recte* Vaginulininae Reuss, 1860, nom. transl. Reuss, 1862, ex family, nom. corr.] included *Planularia, Vaginulina, Falsopalmula, Citharina* and *Citharinella*.

Various lineages have been suggested for the development of the rectilinear Nodosariidae from the Nodosinellid ancestry (e.g., *Lunucammina* to *Ichthyolaria* to *Frondicularia*; *Prodentalina* to *Mesodentalina* to *Dentalina*; *Nodosinella* to *Nodosaria*), but others have yet to be restudied. Probably additional generic taxa will be reinstated that have been previously regarded as synonyms, as were *Ichthyolaria, Lingulinella* and others. In all such studies, strict adherence to the ICZN is necessary. This distinctly forbids substitution of new "type species", as suggested by Civrieux and Dessauvagie (1965) for *Lingulonodosaria, Colaniella*, and *Pachyphloia* (see ICZN Arts. 61 (paragraph 1), 67 i (ii), 68 a); because *Frondinodosaria* Civrieux and Dessauvagie, 1965, included the type species of *Lingulonodosaria*, it is a junior synonym of the latter.

The Nodosariacea is regarded as having its origin in the Endothyracea, but evidence that the rectilinear nodosariids arose from the Nodosinellidae, whereas the enrolled or arcuate genera had their origin in the Loeblichiidae, makes advisable a subdivision of the Nodosariidae of the Treatise and earlier publications. The Nodosariidae as restricted by Putrya (1970) includes genera that arose from various genera within the Nodosinellidae, as discussed earlier. In view of the above, and on the basis of priority of family nomenclature, the Family Vaginulinidae Reuss, 1860, is reinstated herein (=Lenticulinidae auctt.), with subfamilies Lenticulininae Chapman, Parr and Collins, 1934 (=Lenticulininae, Astacolinae Putrya), Marginulininae Wedekind, 1937 (=Vaginulinopsinae Putrya), and Vaginulininae Reuss, 1860 (=Planulariinae Putrya).

The genus *Pseudarcella* Spandel, 1909, was included in the Nodosariinae in the Treatise, but thin section studies of the imperforate wall, and scanning electron microscopy of the exterior and interior surfaces of various species resulted in its transfer to the Tintinnida and removal from the Foraminiferida (Tappan and Loeblich, 1968), together with other related genera of the Eocene and Oligocene, *Remanellina, Tytthocorys* and *Yvonniellina*, all of Tappan and Loeblich, 1968 (the last-named genus including as synonym *Conicarcella* Keij, 1969), *Spinosphenia* and *Urnulella*, both of Szczechura, 1969, *Spinarcella* Keij, 1969, and *Nipterula* Farinacci, 1969.

Among other new genera proposed for the Nodosariidae are *Sieberina* Fuchs, 1970, *Neogeinitzina* Brotzen, 1963 (non K. V. Miklukho-Maklay, 1954) which was renamed *Geinitzinita* Civrieux and Dessauvagie, 1965 (also as *Paralingulina* Gerke, 1969, which is thus a junior synonym); *Pseudolangella* and *Frondina*, both of Civrieux and Dessauvagie, 1965, the former considered a synonym of *Pseudonodosaria* by Fuchs, 1970; *Pachyphloides* Civrieux and Dessauvagie (the type species *P. oberhauseri* is a superfluous synonym of *Lingulina infirmis* Oberhauser, and should be cited as *P. infirmis* (Oberhauser); this genus probably is a junior synonym of *Pseudofrondicularia* Wedekind, although restudy of the type species of the latter may be necessary for confirmation), *Langella* Civrieux and Dessauvagie (new name for *Padangia* Lange, 1925, non Babor, 1900, nec Werner, 1924; reinstated as distinct from *Geinitzina*), *Sosninella* Civrieux and Dessauvagie (based on drawings of random sections published by Sosnina for which no specific name was given, hence invalid ICZN Art. 13b); "*Howchinella*" Civrieux and Dessauvagie is of questionable validity (ICZN Arts. 13 (b), 67 (c)), as *Frondicularia woodwardi* Howchin is stated to be the type species on p. 118, but *Geinitzina caseyi* Crespin is so designated on p. 172; the authors also state that information as to the internal structure is necessary before a diagnosis is possible; *Tollmannia* Civrieux and Dessauvagie is based on figures of the Miocene *Lingulina costata tricarinata* Tollmann, for which also no internal data is available, and which may equally well be retained in *Lingulina*, as it occurs with typical *L. costata*, and may represent merely an occasional aberrant specimen or at most a subspecies; four other Permian genera of Civrieux and Dessauvagie, 1965, are each based on single species from the same locality, *Cryptoseptida* and *Cryptomorphina* are possibly individual variations of *Langella*, and *Calvezina* and *Tauridia* variants of *Lunucammina*.

Other taxonomic changes suggested by Civrieux and Dessauvagie (1965) are not valid: *Lunucammina* is the correct name for the taxon they term *Geinitzina*, as stated in the Treatise (ICZN Arts. 23e (iii), 44b, 60a) even if the type species is based on a deformed specimen. They reinstate *Falsopalmula*, stating that *Palmula* (with Paleocene type species) is unrecognizable. However, the original types of the type species of *Palmula* in the collections of the Philadelphia Academy of Natural Sciences, were redescribed by Howe (1936), and the species was figured in the Treatise from the type area. Whether the microstructure of the Jurassic *Falsopalmula* differs from that of *Palmula* s.s. sufficiently to warrant separation has not been determined. They similarly state that *Nodosinella* is unrecognizable, and thus reinstate *Protonodosaria* which was regarded as a synonym in the Treatise. Like all of Brady's Carboniferous and Permian species, *Nodosinella* was based upon entire specimens, the lectotype of *N. digitata* from the English Permian, in the British Museum (Natural History), being redrawn for the Treatise; a sectioned paratype was figured by Cummings (1955) to elucidate the internal structure. However, the Permian *Protonodosaria* has not been demonstrated as sufficiently distinct for separation.

Among the Polymorphinidae, new genera include the Lower Cretaceous *Pseudopyrulinoides* Fuchs, in Fuchs and Stradner, 1967 in the Polymorphininae, and *Edithaella*, *Cornusphaera*, and *Grillita*, all Fuchs, 1967, and all in the new subfamily Edithaellinae, and *Echinoporina* Fuchs, 1967, in the Webbinellinae.

2. SUPERFAMILY SPIRILLINACEA

Very little modification has involved this superfamily; it includes simple, non-septate enrolled tubular, or biserially formed taxa. Unlike other hyaline calcareous taxa with distinct preferred orientation of the calcite crystals, the Spirillinacean test appears to consist optically of a single crystal of calcite.

The new subfamily Ungulatellinae Seiglie, 1964a, was proposed as intermediate between the Spirillininae and Patellininae, including *Ungulatella*, *Patellinella* and a new genus *Ungulatelloides*. Of these, *Patellinella* had been placed in the Discorbinae in the Treatise, in view of the perforate radial wall and lack of a nonseptate early spire, and later was transferred to the subfamily Conorbininae (Tappan and Loeblich, 1966); *Ungulatella* was placed in the Robertinidae in the Treatise, and tentatively *Ungulatelloides* also seems best placed therein, pending X-ray examination of the test to determine if it is in fact aragonitic.

The family Involutinidae was tentatively placed in the Cassidulinacea in the Treatise, on the basis of the lamellar microgranular wall, although specimens are generally poorly preserved and recrystallized. Oberhauser (1964) restudied *Trocholina* from the Middle and Upper Triassic, and placed *Trocholina* and *Triasina* with *Permodiscus* in the mainly Paleozoic Archaediscidae. *Aulotortus* and *Paratrocholina* were regarded as synonyms of *Permodiscus*. In another study, the microgranular wall of this group was regarded as due to recrystallization (Koehn-Zaninetti, 1969), and the family related to the Spirillinidae. *Aulotortus, Paratrocholina, Angulodiscorbis,* and *Rakusia* Salaj, 1967, were placed in the synonymy of *Involutina*. A family Trocholinidae Kristan-Tollmann, 1963 was regarded as having developed independently, but in parallel with the Archaediscidae and the Cornuspirinae from an ammodiscid ancestor. The Trocholinidae was placed in the synonymy of the Involutinidae (Loeblich and Tappan, 1964c).

3. SUPERFAMILY DUOSTOMINACEA BROTZEN, 1963

(Nom. transl. herein, ex family Duostominidae Brotzen, 1963).

Test free, enrolled, planispiral to high trochospiral, wall non-lamellar, fibrous to granular, and may incorporate some foreign matter; aperture single or double, interiomarginal.

Included in this superfamily are the families Asymmetrinidae Brotzen, 1963, Duostominidae Brotzen, 1963 (synonym: Variostomatidae Kristan-Tollmann, 1963), and Oberhauserellidae Fuchs, 1970.

The origin of the hyaline lamellar foraminifera has been obscured by the absence of well-preserved good transitional faunas of Permo-Triassic age. The hyaline foraminifera have been variously regarded as derived from the Endothyracea (Loeblich and Tappan, 1964b), Lituolacea (Glaessner, 1963), or both (Brotzen, 1963). Because of the much greater abundance of the Nodosariacea at this time, their evolutionary sequence is becoming relatively certain, but in contrast, some of the Triassic genera of the three families here included are known only from a single specimen (*Asymmetrina, Plagiostomella*).

Brotzen (1963) erected the families Asymmetrinidae and Duostominidae as ancestral to the Rotaliidea, whose evolution from an agglutinated or calcareous microgranular wall to a hyaline lamellar one paralleled the evolution from the Endothyracea to the Nodosariacea. Most of the Triassic genera were found to have multiple or lobate apertures, a feature retained by the aragonitic Ceratobuliminidae and Epistomininae (Brotzen, 1963).

The third family of primitive Mid-Triassic to Lower Jurassic genera, the Oberhauserellidae (Fuchs, 1970) includes the genera *Kollmannita, Schmidtia, Oberhauserella, Schlagerina*, and *Praegubkinella* (all of Fuchs, 1967). Representatives of this group have been known as the "Jurassic Globigerinids", variously regarded as true *Globigerina*, as agglutinated pseudomorphs, or ancestral types. Regarded as derived from *Diplotremina* (Duostominidae), they were considered (Fuchs, 1970) as ancestral variously to the trochospiral planktonic Hedbergellinae, *Gubkinella* and the Guembelitriidae, to the Ceratobulimininae and Epistomininae, and to the Discorbinae. Of these included genera, *Praegubkinella* was reported to be non-lamellar, with an imperforate wall of fibrous radiate aragonite. Detailed information is not available for most of these Triassic genera as to wall composition and structure although such information is necessary for corroboration of the suggested evolutionary lineages.

4. SUPERFAMILY ROBERTINACEA

Characterized by the aragonitic, hyaline, radial perforate test, two families were recognized in the Treatise, the Ceratobuliminidae (with subfamilies Ceratobulimininae and Epistomininae), and Robertinidae. McGowran (1966a) divided the Robertinidae into three groups, the subfamilies Ceratobulimininae (placed here rather than with the Epistomininae, following Brotzen), Robertininae and a new subfamily Alliatininae. Although these three groups include closely related genera, as defined, less than half of the genera were allocated to any group, and about one half of the evolutionary derivations were questioned.

Srinivasan (1966) described a new Late Eocene genus of the Ceratobulimininae, *Vellaena*, which was regarded as intermediate between *Ceratobulimina* and *Stomatorbina* of the Epistomininae, corroborating the close relationship of these subfamilies. In a still different interpretation, Reiss (1963) proposed a subfamily Conorboidinae (ex family Conorboididae Hofker, 1951), in which he included the genera *Conorboides* (Ceratobulimininae), *Reinholdella* (Epistomininae) and *Colomia* (Robertinidae)! A new subfamily Reinholdellinae Seiglie and Bermúdez (1965a) was proposed to include three other genera, *Reinholdella*, *Garantella* and *Rubratella*, of which the first two had been placed in the Epistomininae and the other in the Ceratobulimininae. Restudy of many more of the included genera must precede any major reclassification.

5. SUPERFAMILY BULIMINACEA

The high trochospiral, perforate radial hyaline foraminifera, with later biserial and uniserial modifications and characterized by a slit-like aperture, were placed in the Buliminacea, whereas perforate granular taxa with convergent morphology were placed in the Cassidulinacea. Those with most distinctive characteristics and ornamentation (*Uvigerina*, *Sagrina*, *Millettia*, *Siphogenerinoides*) are invariably radial and thus included in the Buliminacea, whereas the apparent pseudomorphs that are granular-walled invariably have a generalized form and lack distinctive apertural or surficial modifications. Thus, convergent morphology does not yet appear to be ruled out.

Ancestry of the Buliminacea is uncertain. The geologically oldest certain representative is the Liassic *Brizalina*, which according to Nørvang (1957) is built of fibrous aragonite. Although *"Bolivina"* (=*Brizalina*) has been reported from the Late Triassic, it is there represented only by a steinkern, and possibly could be referred to *Cassidella* instead, or even to a loosely cemented agglutinated foraminifer. There is also no available data concerning the possible presence of an internal toothplate in any Liassic species. Although the origin of the group remains uncertain, Glaessner (1963) suggested that all other hyaline lamellar forms had an origin independent of that of the Nodosariidae. If so, the Buliminacea may have had their ancestry among such calcareous microgranular Paleozoic Endothyracea as *Deckerellina* (Palaeotextulariidae), parallelling the Nodosariid origin from the Nodosinellidae; evolution from the agglutinated Textulariidae seems less likely.

No major reclassifications have been made within this superfamily, and the family groups proposed appear to be unnecessary. Thus, the Chiloguembelinidae Reiss, 1963, was proposed on the basis of their supposedly monolamellar wall, in contrast to the bilamellar one of the true Heterohelicidae. Since that time, all "monolamellar" hyaline taxa have been shown to be in fact of bilamellar construction. The subfamily Schubertiinae Reiss, 1963 (based on the homonym *Schubertia, recte Millettia*) included only two genera of the Bolivinidae which had partial subdivisions of the chambers, and the Stainforthiinae Reiss, 1963 included three genera, but was placed in the Virgulinidae (*recte* Caucasinidae), a granular-walled group. The Trifarininae Srinivasan, 1966 (family Uvigerinidae) was proposed to include triangular tests, with triserial to uniserial construction and terminal aperture.

However, reinstatement of the subfamily Buliminellinae Hofker, 1951 (nom. transl. Bykova in Rauzer-Chernousova and Fursenko, 1959, ex family) appears worthwhile, to include only those taxa with many broad, low chambers per whorl (*Buliminella, Buliminellita, Buliminoides*, and *Quadratobuliminella*), typically of Cenozoic age, but perhaps including Late Cretaceous representatives (many older *Buliminella* have been transferred to *Praebulimina, Aeolostreptis*, etc.).

6. SUPERFAMILY DISCORBACEA

Recent revisions of the families and genera of the Discorbacea have involved many aspects of the wall microstructure, septal lamellar character and apertural features, some of which are briefly summarized below. Reiss (1963) proposed many new family group taxa, in part based on septal lamellar character (before recognition by means of electron microscopy of the similarity of the so-called monolamellar and bilamellar structure). New taxa were the subfamily Asterigerinatinae (originally placed in the aragonitic Conorboididae; this subfamily was elevated to family status by Singh and Kalia, 1972), the subfamily Conorbininae (recte nom. transl. ex family of Hofker, 1954; also originally referred to the Conorboididae, but transferred to the Discorbidae by Tappan and Loeblich, 1966); the Epistominellinae (synonym of Pseudoparrellidae Subbotina, 1959, which therefore was reinstated and expanded by Lipps, 1965); the Eponidopsidae (=Eponididae Hofker, 1951, their type species being synonymous); Heminwayininae (in the family Rosalinidae, but the type genus was shown in the Treatise to be a synonym of *Eoeponidella*; Seiglie, 1964a, p. 511; 1964b, p. 3; and Seiglie and Bermúdez, 1965a, p. 159, concurred in this synonymy and similarly placed *Asterellina* Andersen, 1963, in synonymy of *Eoeponidella*); the Rosalinidae and Rosalininae (the family was restricted by Loeblich and Tappan, 1964c to include only *Rosalina* and the synonym *Tretomphalus*, and *Eoeponidella*, but it is better regarded as a subfamily of the Discorbidae).

Other family group taxa are the monogeneric Alfredininae Singh and Kalia, 1972 in the Asterigerinidae (the type genus may be a synonym of *Epistomaroides*); Cribroeponidinae Shschedrina, 1964 (for the multiple-apertured Eponididae, although others do not regard these as even of generic significance); the Eponidellinae Seiglie and Bermúdez, 1965 (a subdivision of the Epistomariidae to include *Eponidella, Palmerinella*, and *Epistomaroides*), the radial walled Planulinidae and superfamily Planulinacea both of Gonzàles-Donoso, 1969 (ex subfamily of Bermúdez, 1952), and the granular-walled Heterolepidae Gonzàles-Donoso, 1969, both regarded as having a pseudo-trilamellar septal structure rather than bilamellar (but in fact showing the typical inner lining, median layer and outer lamella elucidated from electron microscopy by Hansen and Reiss, 1971, 1972, and thus not valid distinctions); and the monogeneric Serovaininae Sliter, 1968, in the Discorbidae (radial, monolamellar, and similar to the Baggininae).

In a partial restudy of the Discorbacea (excluding only the Asterigerinidae, Glabratellidae and Siphoninidae), Tappan and Loeblich (1966) included four families, the Discorbidae (with subfamilies Conorbininae, Baggininae, Discorbinae; various genera were redefined by Douglas and Sliter, 1965); Pseudoparrellidae (as modified by Lipps, 1965 to include the reinstated *Pseudoparrella, Epistominella, Stetsonia, Megastomella* Faulkner *et al.*, 1963, and the new genera *Ambitropus* and *Concavella*); the reinstated Laticarininidae (for *Laticarinina, Biapertorbis, Discorbinella, Planulinoides* and *Torresina*) and the Epistomariidae (with the addition of *Helenina*). Excluded from the Discorbacea were *Aboudaragina* (transferred to the Alabaminidae), *Buccella* and *Vernonina* (placed in the Rotaliidae), *Rosalina, Earlmyersia*, and *Heminwayina* (placed in the Rosalinidae), and *Discorbitura* (in the Anomalinidae). The Triassic *Variostoma, Duostomina*, and *Diplotremina* had already been removed to the new Duostominidae by Brotzen, 1963.

Some of the Discorbacean families, including the Glabratellidae were restudied by Seiglie and Bermúdez (1965a, b). The Glabratellidae included *Glabratella, Pijpersia, Pseudoruttenia* (reinstated as distinct), *Pileolina, Heronallenia, Angulodiscorbis*, and the new genera *Heronallenita, Neoglabratella, Planoglabratella, Glabratellina, Corrugatella, Fastigiella*, and *Claudostriatella*. They also included the granular-walled *Trichohyalus*, however.

The Pegidiidae was also reinstated as a family and the subfamily Rupertininae transferred to the Cibicididae (by Loeblich and Tappan, 1964c).

As the electron microscopy data now indicates no true distinction between monolamellar and bilamellar taxa, certain families that had been included in the Orbitoidacea seem best transferred to the Discorbacea, restricting the former to the families Orbitoididae, Discocyclinidae, Lepidocyclinidae, and Pseudorbitoididae.

The family Indicolidae Singh and Kalia, 1970, originally was placed in the Globigerinacea, but of the two eventually included genera, *Indicola* Singh and Kalia, 1970, from the Lutetian, strongly resembles the similarly Eocene *Eorupertia*, and is better placed in the Victoriellinae, and *Praeindicola* Singh and Kalia, 1971, resembles the similarly Eocene *Neocribrella*; El-Naggar (1971, p. 476) suggested that *Indicola* and *Praeindicola* were assignable to the Eponididae, and the latter was a probable synonym of *Cincoriola*.

7. SUPERFAMILY GLOBIGERINACEA

Many of the discussions of wall structure and septal lamellar character have involved the planktonic taxa wholly or in part. The planktonic foraminifera have generally been regarded as having radial perforate hyaline walls but Pessagno (1964) reported that septa of all planktonic species examined were microgranular in texture, the wall itself was macrogranular and the ornament areas ultragranular in texture. Similar granular texture was reported in septa of *Globorotalia* and *Pulleniatina* (by Hansen *et al.*, 1969), but these were also reported to be optically radial. The walls similarly were stated to be not truly radial (Blow, 1969), but to consist of microgranular crystals stacked within protein membranes, the cross struts of protein resulting in a pseudo-lamellar appearance, not comparable to the lamellation resulting from periodic growth. Blow (1969) reported further that the preferential orientation of the granules resulted in the optical extinction pattern in polarized light. Such a structure later was termed "optically radiate, ultrastructurally nonradiate" (Hansen, 1970b).

The planktonic septal wall was variously reported as monolamellar, nonlamellar, and bilamellar in numerous articles, at times on the evidence supplied by the same photographs. Bé and Hemleben (1970) state that the wall is bilamellar, but composite, with many individual units in each lamella, as had been reported also by Blow (1969). An inner lamellar unit consists of anhedral (microgranular) calcite; a primary organic membrane separates this layer from an outer lamellar unit which consists of many lamellae of anhedral to euhedral crystals. In a later phase of growth, a calcite crust of euhedral crystals is added to the exterior, and in some (*Sphaeroidinella*), a final amorphous veneer or cortex was added to the surface (Bé and Hemleben, 1970). Much of the conflict in interpretation was apparently due to varying use of terminology. Reiss and Luz (1970) state that the planktonics are bilamellar; each primary chamber wall, secreted during one episode of skeletal growth (instar), consists of an inner calcareous lining confined to a single chamber, an outer calcareous lamella which continues over the exterior of earlier chambers, and an organic (proteinaceous) median layer. The latter consists of a three-dimensional network which occasionally incorporates some calcite crystals, and stops abruptly at the contact with the previous septum. All planktonic taxa are bilamellar, although the median layer may be extremely thin and poorly defined (Hansen and

Reiss, 1972), but even then the pore tubules may be constricted at its position in the wall. Similarly, all species show both the primary layering (thin, multiple lamellae, the individual units of others), and the secondary lamination due to episodic growth.

Various bases have also been used in modification of the classification of the planktonic taxa, generally including only the Cretaceous taxa (Maslakova, 1964; Pessagno, 1967; Fuchs, 1971), or Cenozoic ones (Lipps, 1964, 1966; Parker, 1967; McGowran, 1968; Steineck, 1971; Subbotina, 1971), although some include both (Tappan and Lipps, 1966; El-Naggar, 1971).

For the Cretaceous groups, Maslakova (1964) subdivided the Globotruncanidae into the Globotruncaninae, Rotaliporinae, Rugoglobigerininae, and the new subfamily Globotruncanellinae (for *Globotruncanella* and *Globotruncanita*, both regarded as synonyms of *Globotruncana* in the Treatise). Pessagno (1967) used a classification similar to that of the Treatise, but erected a new monotypic subfamily Loeblichellinae in the Rotaliporidae, and divided the Globotruncanidae into the Globotruncanidae with *Globotruncana*, *Rugoglobigerina*, *Rugotruncana* and the new *Archaeoglobigerina*; the new family Marginotruncanidae, including *Marginotruncana* and the new *Whiteinella*, and the new family Abathomphalidae, containing *Abathomphalus* and *Globotruncanella* and thus preoccupied by the Globotruncanellinae Maslakova, 1964. However, Van Hinte (1963, p. 96) regarded *Globotruncanella* as a synonym of *Abathomphalus*, and Douglas (1969) placed *Whiteinella* and *Loeblichella* in the synonymy of *Hedbergella*. Fuchs (1971) combined drastically the family groups of Cretaceous planktonics. He recognized two families, the Guembelitriidae, including the new genus *Iuliusina*, and the Hedbergellidae (elevated from subfamily rank, but including the Hedbergellinae, Heterohelicinae, Rotaliporinae and Globotruncaninae, hence preoccupied by all other included named groups, and correctly should have been termed the Heterohelicidae Cushman, 1927; in addition, the subfamily Hedbergellinae as recognized by Fuchs included genera of the former Planomalinidae and Schackoinidae, and thus should have been termed Planomalininae Bolli, Loeblich and Tappan, 1957).

On the basis of wall microstructure and surface texture, Lipps (1964, 1966) separated the Cenozoic planktonic foraminifera into those with smooth surface, constructed of identical nearly parallel crystals oriented perpendicular to the test surface, the Hantkeninidae (which included the Hantkenininae and Cassigerinellinae), and Globorotaliidae (including in 1964 the Globorotaliinae, Catapsydracinae and Globigerinitinae, modified in 1966 to include only the Globorotaliinae and Candeininae); the spiny Globigerinidae in which some crystals are elongated as spines,

and are surrounded by finer crystals (including the subfamilies Globigerininae, Orbulininae and Hastigerininae); and the group with pitted
surface, and wall consisting of short prominent crystals surrounded by
finer ones, which was believed ancestral to the spiny type and later
elevated to family rank as the Catapsydracidae (Tappan and Lipps,
1966; Lipps, 1966). Lipps suggested that many of the morphologic
adaptations of the planktonic foraminifera were related to problems of
flotation rather than indicating phylogeny and that many described
"lineages" merely indicated successive occurrences in a local area; thus
the great similarity of the juvenile specimens and early stages resulted
from the similar stresses and convergence, rather than being applicable
to the determination of relationships.

Parker (1967) recognized the same four families as in Lipps, but
elevated the Candeinidae to family status, regarding this group as
distinct from the Globorotaliidae; this group may have short, fine
secondary spines on the surface. Some Neogene genera were regarded
as polyphyletic (e.g., *Globigerinoides*, *Globoquadrina*, and *Globorotalia*),
Globoquadrina was emended to include many species previously placed
elsewhere, but which similarly were trochoid with pitted wall surface.
Many lineages of *Globorotalia* were said to evolve from rounded periphery
to keeled, hence *Turborotalia* was regarded as a synonym; *Globorotaloides*
was placed in the synonymy of *Globoquadrina*.

A new technique for determination of relationships among the
planktonic species was suggested by King and Hare (1972). Amino acid
analyses were made of 16 recent species, each proving to have a unique
composition, but these varied in degree. In a Q-mode factor analysis of
their amino acid composition, non-spinose species were separated from
spinose ones, as has classically been done in the Globorotaliidae and
Globigerinidae. *Pulleniatina* and *Globoquadrina* were close in composition
(these representing the Catapsydracidae). The greatest variation was
shown in the spinose species, for example in *Globigerinoides*, agreeing with
Parker's suggestion of its polyphyletic origin. Most *Globigerinoides* species
were similar to *Globigerina*, but *Globigerinoides sacculifer* and *G. conglobatus*
were close to *Sphaeroidinella dehiscens*, as previously had been suggested
from other data (Bé, 1965; Bé and Hemleben, 1970; and Parker, 1967).
Two Miocene *Globoquadrina* were analyzed and found to be closely
similar to the living *G. dutertrei* in composition. The unusually good
evidence provided by this method suggests that it will prove useful in
the future for determination of relationships.

In other modifications of the classification of Cenozoic planktonic
foraminifera, McGowran (1968) and Steineck (1971) placed in the
family Globigerinidae the subfamilies Globorotaliinae, Truncorotaloidinae, Catapsydracinae, and Globigerininae; Subbotina (1971)
recognized three families, the Globigerinidae, Catapsydracidae and

Globorotaliidae, the latter including the Globorotaliinae and Trunco-rotaloidinae, and two new subfamilies, the Acarininae (recte Acarinini-nae) and Truncorotaliinae.

In a general reclassification of the planktonic foraminifera, El-Naggar (1971) discounted the differences in wall structure and emphasized gross morphology, chamber arrangement, and apertural modifications at the family level, his classification including 5 families and 6 subfamilies. The Guembelitriidae was elevated from subfamily status (including *Gubkinella* and *Globigerina (Conoglobigerina)* as synonyms of the type genus, *Guembelitriella*, and with the new genus *Guembeli-trioides*); the Globotruncanidae, subfamily Globotruncaninae, including *Rugoglobigerina* (with *Archaeoglobigerina* as a subgenus, and including *Whiteinella* and probably *Globigerina (Globuligerina)* Bignot and Guyader, 1971, as synonyms), *Globotruncana*, and subgenus *Rugotruncana*, *Plummerita* and new subgenus *Radotruncana*, subfamily Rotaliporinae (including *Rotalipora, Hedbergella, Ticinella, Praeglobotruncana, Clavihedbergella* and the new *Claviticinella*), subfamily Planomalininae (including *Globigerinel-loides, Biticinella, Planomalina, Schackoina, S. (Hastigerinoides)*, and *S. (Eohastigerinella)*), Globigerinidae, with subfamily Globigerininae *(Globigerina, Globigerinoides, Globoquadrina, Schackoinella, S. (Beella)*, the new *Globigerinoidesella, Candeina, Sphaeroidinellopsis, Sphaeroidinella, Globiger-inopsis, Globigerapsis, Globigerinatheka, Inordinatosphaera* Mohan and Soodan, 1967, *Globigerinatella, Orbulina*, and *Candorbulina*), subfamily Globorotaliinae *(Turborotalia, Globorotalia, Truncorotaloides, Astrorotalia, A. (Clavatorella), Pulleniatina, Globigerinita, Orbulinoides* Blow and Saito, 1968), subfamily Hantkenininae *(Hastigerina, Globanomalina, Hantkenina, H. (Bolliella), H. (Clavigerinella), Cribrohantkenina, Cassigerinella, Cassiger-inelloita* Stolk, 1965, *Hastigerinella*), family Heterohelicidae *(Heterohelix, Pseudoguembelina, Gublerina, Ventilabrella, Lunatriella* Eicher and Worstell, 1970), family Chiloguembelinidae Reiss, 1963 *(Chiloguembelina*, the new genus *Chiloguembelinella*).

8. SUPERFAMILY ROTALIACEA

Changes in the classification of the Rotaliacea have been minor, although affected by the controversy as to the taxonomic value of the radial and granular microstructure, for example, in separating *Elphidium* and the Nonionidae. Studies of the wall structure and lamination in *Ammonia* (Hansen, 1970b, Hansen and Reiss, 1971) were reviewed herein in the general discussion of the Rotaliina. Other evidence from electron microscopy has elucidated surface texture, porosity, and wall structure in the Elphidiidae (Hay *et al.*, 1963; Lynts and Pfister, 1967). Reiss (1963) erected a subfamily Pararotaliinae within the Miscellaneidae, but most of the included genera were placed in the Cuvillierininae in the Treatise, whereas *Pararotalia* was included in the Rotaliinae.

9. SUPERFAMILY ORBITOIDACEA

As mentioned before, the elimination of monolamellar or bilamellar septa as a major character for classification, suggested by the detailed information from electron microscopy supports the transfer of the less complex foraminiferal families from the Orbitoidacea to the Discorbacea, with the retention only of the Orbitoididae, Discocyclinidae, Lepidocyclinidae and Pseudorbitoididae.

MacGillavry (1963) again separated the Lepidorbitoididae and Orbitoididae, the former containing *Lepidorbitoides, Actinosiphon, Asterorbis*, the reinstated *Orbitocyclina*, and *Orbitocyclinoides*, and the new genus *Helicorbitoides*. The Pseudorbitoididae was subdivided into two subfamilies, the Pseudorbitoidinae including *Pseudorbitoides* and *Sulcorbitoides*, and three genera removed from synonymy as distinct, *Conorbitoides, Historbitoides*, and *Rhabdorbitoides*; the new subfamily Vaughanininae included *Vaughanina, Ctenorbitoides* and the reinstated *Aktinorbitoides*.

Caudri (1972) suggested that Brönnimann's separation of the forms included in the Discocyclinidae into two distinct groups should be maintained at the family level. She thus recognized the Discocyclinidae as including *Discocyclina* s.s. (restricted to the Eastern Hemisphere), *Aktinocyclina, Proporocyclina, Athecocyclina*, and *Asterophragmina*, whereas the reinstated Orbitoclypeidae contained *Orbitoclypeus*, two new genera, *Neodiscocyclina*, and *Stenocyclina, Pseudophragmina* s.s., *Asterocyclina*, and probably *Hexagonocyclina*.

In contrast to these subdivisions of the family groups, Neumann (1972) considered that the differences between them were not of family rank, and combined the Discocyclinidae and Lepidocyclinidae in the single family Orbitoididae. The median black line in their wall was determined not to represent a canal system, but a median organic layer similar to that in the wall of the smaller hyaline foraminifera.

10. SUPERFAMILY CASSIDULINACEA

As recognized in the Treatise, the Cassidulinacea included the entire spectrum of microgranular lamellar hyaline foraminifera; it was later divided (Loeblich and Tappan, 1964c) into three superfamilies. The restricted Cassidulinacea included those forms with basically high trochospiral to biserial chamber arrangement, and slit-like or loop shaped aperture, commonly with internal tube or toothplate. Included in this superfamily were the Pleurostomellidae, Annulopatellinidae, Caucasinidae, Delosinidae, Loxostomatidae, and Cassidulinidae. A new subfamily, Cribropleurostomellinae Owen, 1971, was proposed for the Late Cretaceous cribrate *Cribropleurostomella* Owen, 1971; as the latter genus seems closely related to other genera of the family, particularly *Pinaria*, which also has a multiple aperture, a separate subfamily seems unnecessary.

The monotypic Tremachoridae Lipps and Lipps, 1967, was erected to include a Miocene trochospirally enrolled form with areal slit-like aperture and sutural supplementary apertures, that possibly is ancestral to the Delosinidae.

11. SUPERFAMILY NONIONACEA

The superfamily Nonionacea was described as including the mono-lamellar enrolled perforate-granular hyaline foraminifera and the superfamily Anomalinacea included the bilamellar taxa (Loeblich and Tappan, 1964c). As the lamellar character now appears an invalid basis for separation, the latter superfamily now is combined with the Nonionacea. Furthermore, the Involutinidae was transferred to the Spirillinacea from the Nonionacea, as discussed previously, as the wall is recrystallized, and not comparable to that of the true perforate granular taxa. Similarly, the Triassic taxa, *Asymmetrina*, *Involvina*, and *Plagiostomella* are also removed.

The family Pulleniidae Putrya, 1963, (recte nom. correct. pro Pullenidae Schwager, 1877) was proposed to include *Pullenia*, *Cribropullenia*, and *Allomorphinella*; the first two were placed in the Nonionidae in the Treatise, and the last was in the Chilostomellinae.

The family Heterolepidae Gonzàles-Donoso, 1969, was proposed on the basis that *Heterolepa* has a trilamellar wall, as previously mentioned. The nature of the wall has been further elucidated since then (Hansen and Reiss, 1971) and is regarded as closely similar to that of other Anomalinidae.

As now recognized herein, the Nonionacea consists of the families Nonionidae, Alabaminidae, Osangulariidae and Anomalinidae.

12. SUPERFAMILY CARTERINACEA

The single included genus, *Carterina*, remains the only foraminifer to have a wall of secreted calcite monocrystalline spicules. The delicate test disintegrates soon after death, hence only living species are known.

IV. Phylogeny at the Family Level

The expanding data base for the Foraminiferida is well indicated by the degree of familial classification, both of which have continued to increase. As briefly discussed in the preceding pages, many of the recently proposed family-group categories do not appear useful. Obviously, in any family or subfamily, certain genera are closely related, and others less closely so, but not all of these relationships require formal classificatory recognition. Similarly, various phylogenies have been proposed for different family groups, as in the Lituolacea, Fusulinacea, Robertinidae, or the planktonic genera, and family classification modified to correspond; unfortunately, many of the phylogenies thus proposed utilize only the selected genera which do show close relationship, and ignore others, without comment as to whether they should be placed elsewhere, or retained and eventually located within the phylogenetic scheme. In the accompanying Fig., phylogenetic relationships are suggested, with the inclusion of every family recognized herein for the Foraminiferida. Subfamilies are not indicated, in view of the scale required.

In a phylogenetic classification, any higher, or more recently evolved group should arise from a single predecessor or predecessor group of the same or lower rank (Corliss, 1972). Therefore, when a descendent taxon as presently conceived is regarded as having diverse predecessors of the same rank, the polyphyletic situation should be revised. Polyphyletic origins involving diverse groups of lower rank are legitimate; although a family must have its origin in a single other family, it may include subfamilies (or genera) that arose from different subfamilies (or genera, respectively) of the ancestral family.

The present classification agrees with these requirements in nearly every respect. Thus, the Lituolacea is regarded as derived from the Ammodiscacea, although the Hormosinidae is suggested to have been derived from the Astrorhizidae (probably the Hippocrepininae) and the Rzehakinidae from the Ammodiscidae (Ammodiscinae). Similarly, the Nodosariacea arose from the Endothyracea, but the Nodosariidae from the Nodosinellidae (Nodosinellinae) and the Vaginulinidae from the Loeblichiidae. In most other instances, a single family is regarded as ancestral to the next family, families, or superfamily, although different genera may have separate origins within the ancestral family. The origins of the Cassidulinacea are more problematic. Apparently, the Nonionacea arose from the Asymmetrinidae, and in turn may have given rise to the Caucasinidae. However, it is difficult to see any relation to the Pleurostomellidae other than the character of the wall. The latter appears similar in many respects to the Nodosariacea, such as the Glandulinidae, in similarity of chamber arrangements, as in *Nodosarella*, *Ellipsoglandulina* and *Ellipsopolymorphina*, the presence of an internal tube related to the aperture in all pleurostomellids and all glandulinids and the arched subterminal aperture of the pleurostomellids like that of *Fissurina* and *Parafissurina*. However, the latter are much younger geologically and thus could not be ancestral. On the other hand, the oldest glandulinid (the Lower Cretaceous *Dainita*) has a radiate aperture; the oldest pleurostomellid, the Jurassic *Pleurostomella jurassica* Haeusler was described as having a lateral aperture, although this was not directly visible. Lower Cretaceous species do have typical pleurostomellid apertures. Thus, evidence as to the ancestry of the pleurostomellids

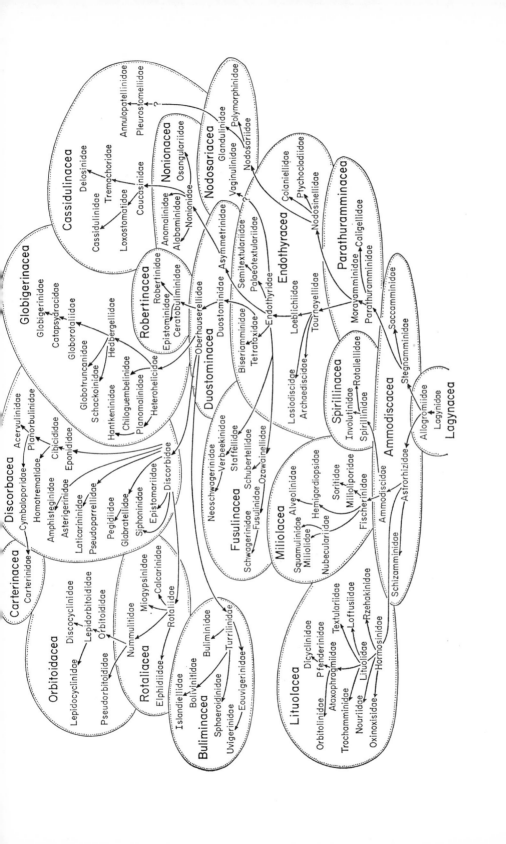

remains equivocal, and as yet insufficient to justify its removal from the Cassidulinacea, whose origins lie within the Asymmetrinidae; as a result, this superfamily remains indicated on the chart as polyphyletic in origin.

The ancestry of the hyaline calcareous foraminifera has become clearer in recent years through the study of the Permian, Triassic and Lower Jurassic assemblages. An excellent sequence has been demonstrated from the Nodosinellidae to Nodosariidae, as already discussed. Origins of the Robertinacea, Discorbacea, Globigerinacea, Buliminacea, and Nonionacea appear to lie within the three families of the Duostominacea, but the extreme rarity of the included genera leaves unanswered many other questions. Perhaps, in part the explanation of the rarity of the primitive taxa may lie in the nature of their test wall. As is well known, aragonite is less stable than calcite, and tends to be selectively dissolved and destroyed. All living aragonitic foraminifera are included within the Robertinacea, and representatives of the Ceratobuliminidae and Epistominidae are present in the Jurassic, and the latter perhaps also in the Triassic. In addition, the oldest *Brizalina* (Buliminacea) from the Lias is aragonitic (Nørvang, 1957), as is the Triassic *Praegubkinella* (Oberhauserellidae), regarded as ancestral to the Globigerinidae (Fuchs, 1969). In the definition of the family Oberhauserellidae, reference was made to the earlier paper on *Praegubkinella*, thus implying an aragonitic wall in all genera. On this basis, as well as the retained double aperture, the Robertinidae appears to be a more primitive group, rather than a later derived one, as we had earlier believed (Loeblich and Tappan, 1964b). Additional well-preserved faunas must be sought in exceptionally fine clay sediments of Triassic age, in which aragonite may be preserved, and may then aid in tracing in detail the early evolutionary diversification of the hyaline calcareous species.

V. Summary of Revised Classification

As modified herein, and previously, we now favor the following classification for the Order Foraminiferida. Changed categories since the Treatise are indicated with an asterisk, and authorship indicated for newly recognized divisions.

Suborder Allogromiina
   Superfamily Lagynacea
      Family Lagynidae
      Family Allogromiidae

Suborder Textulariina
   Superfamily Ammodiscacea
      Family Astrorhizidae
         Subfamily Astrorhizinae
         Subfamily Rhizammininae

  *Subfamily Hippocrepininae (includes Botellininae)
  *Subfamily Dendrophryninae Vyalov, 1966 (nom. transl. herein
    ex family)
  Subfamily Dendrophryinae
 Family Schizamminidae
 *Family Stegnamminidae Moreman, 1930
 Family Saccamminidae
  Subfamily Psammosphaerinae
  Subfamily Saccammininae
  Subfamily Hemisphaerammininae
  Subfamily Diffusilininae
 Family Ammodiscidae
  *Subfamily Ammovolummininae Chernykh, 1967
  Subfamily Ammodiscinae
  Subfamily Tolypammininae
Superfamily Lituolacea
 *Family Oxinoxisidae Vyalov, 1968
 Family Hormosinidae
  *Subfamily Aschemocellinae Vyalov, 1966 (nom. transl. herein
    ex family)
  Subfamily Aschemonellinae
  *Subfamily Reophacinae Cushman, 1910
  *Subfamily Hormosininae (restricted)
  Subfamily Cribratininae
 Family Nouriidae
 Family Rzehakinidae
 Family Lituolidae
  Subfamily Haplophragmoidinae
  Subfamily Sphaerammininae
  Subfamily Lituolinae
  Subfamily Placopsilininae
  Subfamily Coscinophragmatinae
 Family Barkerinidae
 *Family Loftusiidae Brady, 1884 (nom transl. Lankester, 1885, ex
    subfamily)
  *Subfamily Choffatellinae Maync, 1958
  Subfamily Loftusiinae
  *Subfamily Hemicyclammininae Banner, 1966
  Subfamily Cyclammininae
  Subfamily Spirocyclininae
  Subfamily Mesoendothyrinae Voloshinova, 1958 (nom. transl.
    Banner, 1966, ex family)
 Family Textulariidae
  Subfamily Spiroplectammininae
  Subfamily Textulariinae

Subfamily Pseudobolivininae
Subfamily Plectorecurvoidinae
Subfamily Tawitawiinae
Family Trochamminidae
Subfamily Trochammininae
Subfamily Remaneicinae
Family Ataxophragmiidae
Subfamily Verneuilininae
Subfamily Globotextulariinae
Subfamily Valvulininae
Subfamily Ataxophragmiinae
*Family Textulariellidae Grönhagen and Luterbacher, 1966
*Family Pfenderinidae Smout and Sugden, 1962
Subfamily Pfenderininae
*Subfamily Kurnubiinae Redmond, 1964
*Subfamily Coskinolininae Moullade, 1965 (nom. transl.
Cimerman, 1969, ex family)
Family Dicyclinidae
Subfamily Cyclolininae
Subfamily Dicyclininae
Family Orbitolinidae
*Subfamily Dictyoconinae Moullade, 1965
*Subfamily Orbitolininae

Suborder Fusulinina
Superfamily Parathuramminacea
Family Parathuramminidae
Family Caligellidae
Family Moravamminidae
Subfamily Earlandiinae
Subfamily Moravammininae
Superfamily Endothyracea
Family Nodosinellidae
Subfamily Tuberitininae
Subfamily Nodosinellinae
Family Colaniellidae
*Family Ptychocladiidae (restricted)
Family Palaeotextulariidae
Family Semitextulariidae
Family Tetrataxidae
Family Biseriamminidae
Family Tournayellidae
*Family Loeblichiidae Cummings, 1955 (nom. transl. Rozovskaya,
1963, ex subfamily)

Family Endothyridae
    Subfamily Endothyrinae
    Subfamily Haplophragmellinae
    Subfamily Endothyranopsinae
    Subfamily Bradyininae
Family Archaediscidae
    Subfamily Archaediscinae
    *Subfamily Robuloidinae Reiss, 1963
Family Lasiodiscidae
Superfamily Fusulinacea
Family Ozawainellidae
    Subfamily Ozawaiinellinae
    *Subfamily Pseudostaffellinae Putrya, 1956
Family Staffellidae
    *Subfamily Staffellinae
    *Subfamily Pseudoendothyrinae Mamet, 1970
*Family Schubertellidae Skinner, 1931
    Subfamily Schubertellinae
    *Subfamily Boultoniinae Skinner and Wilde, 1954
Family Fusulinidae
    *Subfamily Fusulinellinae Staff and Wedekind, 1910
    Subfamily Fusulininae
    *Subfamily Eofusulininae Rauzer and Rozovskaya, 1958
*Family Schwagerinidae Dunbar and Henbest, 1930
    Subfamily Schwagerininae
    *Subfamily Pseudoschwagerininae Chzhan, 1963
    *Subfamily Polydiexodininae Miklukho-Maklay, 1953
    *Subfamily Chusenellinae Kahler and Kahler, 1966
Family Verbeekinidae
    Subfamily Verbeekininae
    *Subfamily Misellininae Miklukho-Maklay, 1958
    *Subfamily Pseudodoliolininae Leven, 1963
*Family Neoschwagerinidae Dunbar and Condra, 1928
    Subfamily Neoschwagerininae
    *Subfamily Sumatrininae Kahler and Kahler, 1946
    *Subfamily Thailandininae Toriyama and Kanmera, 1968

Suborder Miliolina
    Superfamily Miliolacea
    Family Fischerinidae
        Subfamily Cyclogyrinae
        Subfamily Fischerininae
        Subfamily Calcivertellinae
    *Family Hemigordiopsidae Nikitina, 1969

Family Nubeculariidae
   Subfamily Nubecularinae
   Subfamily Ophthalmidiinae
   Subfamily Spiroloculininae
   Subfamily Nodobacularinae
   Subfamily Discospirininae
Family Miliolidae
   Subfamily Quinqueloculininae
   Subfamily Miliolinellinae
   Subfamily Miliolinae
   Subfamily Fabulariinae
   Subfamily Tubinellinae
Family Squamulinidae
*Family Milioliporidae Brönnimann and Zaninetti, 1971
Family Soritidae
   Subfamily Peneroplinae
   Subfamily Meandropsininae
   Subfamily Rhapydionininae
   Subfamily Archaiasinae
   Subfamily Soritinae
   Subfamily Keramosphaerinae
Family Alveolinidae

Suborder Rotaliina
  Superfamily Nodosariacea
   Family Nodosariidae
     Subfamily Nodosariinae
     Subfamily Plectofrondiculariinae
     Subfamily Lingulininae
   *Family Vaginulinidae Reuss, 1860
     *Subfamily Lenticulininae Chapman, Parr and Collins, 1934
     *Subfamily Marginulininae Wedekind, 1937 (nom. transl.
       Nørvang, 1957, ex family)
     *Subfamily Vaginulininae Reuss, 1860
   Family Polymorphinidae
     Subfamily Polymorphininae
     *Subfamily Edithaellinae Fuchs, 1967
     Subfamily Webbinellinae
     Subfamily Ramulininae
   Family Glandulinidae
     Subfamily Glandulininae
     Subfamily Seabrookiinae
     Subfamily Oolininae

Superfamily Spirillinacea
  Family Spirillinidae
    Subfamily Spirillininae
    Subfamily Patellininae
  *Family Involutinidae
  Family Rotaliellidae
*Superfamily Duostominacea Brotzen, 1963 (nom. transl. herein ex
    family)
  *Family Asymmetrinidae Brotzen, 1963
  *Family Duostominidae Brotzen, 1963
  *Family Oberhauserellidae Fuchs, 1970
Superfamily Robertinacea
  Family Ceratobuliminidae
  *Family Epistominidae Wedekind, 1937
  Family Robertinidae
    Subfamily Robertininae
    *Subfamily Alliatininae McGowran, 1966
    *Subfamily Ungulatellinae Seiglie, 1964
Superfamily Buliminacea
  Family Turrilinidae
    Subfamily Turrilininae
    Subfamily Lacosteininae
    *Subfamily Buliminellinae Hofker, 1951 (nom. transl. Bykova,
      1959, ex family)
  Family Sphaeroidinidae
  Family Bolivinitidae
  Family Islandiellidae
  Family Eouvigerinidae
  Family Buliminidae
    Subfamily Bulimininae
    Subfamily Pavonininae
  Family Uvigerinidae
Superfamily Discorbacea
  Family Discorbidae
    *Subfamily Conorbininae Hofker, 1954 (nom. transl. Reiss,
      1963, ex family)
    *Subfamily Serovaininae Sliter, 1968
    Subfamily Baggininae
    Subfamily Discorbinae
    *Subfamily Rosalininae Reiss, 1963
  Family Eponididae
  Family Glabratellidae
  *Family Pegidiidae Heron-Allen and Earland, 1928

*Family Pseudoparrellidae Voloshinova, 1952 (nom. transl. Subbotina, 1959, ex subfamily)
*Family Laticarininidae Hofker, 1951 (nom. correct Reiss, 1963)
Family Epistomariidae
Family Siphoninidae
Family Asterigerinidae
Family Amphisteginidae
Family Cibicididae
    Subfamily Planulininae
    Subfamily Cibicidinae
    Subfamily Rupertininae
Family Planorbulinidae
Family Acervulinidae
Family Cymbaloporidae
Family Homotrematidae
    Subfamily Homotrematinae
    Subfamily Victoriellinae
Superfamily Globigerinacea
Family Heterohelicidae
    Subfamily Guembelitriinae
    Subfamily Heterohelicinae
*Family Chiloguembelinidae Reiss, 1963
Family Planomalinidae
Family Schackoinidae
*Family Hedbergellidae Loeblich and Tappan, 1961 (nom. transl. Fuchs, 1971, ex subfamily)
Family Globotruncanidae
    Subfamily Rotaliporinae
    *Subfamily Globotruncanellinae Maslakova, 1964
    Subfamily Globotruncaninae
Family Globorotaliidae
    *Subfamily Candeininae Cushman, 1927
    Subfamily Globorotaliinae
Family Hantkeninidae
    Subfamily Hantkenininae
    Subfamily Cassigerinellinae
*Family Catapsydracidae Bolli, Loeblich and Tappan, 1957 (nom. transl. Tappan and Lipps, 1966, ex subfamily)
Family Globigerinidae
    Subfamily Globigerininae
    Subfamily Hastigerininae
Superfamily Rotaliacea
Family Rotaliidae

Subfamily Rotaliinae
Subfamily Cuvillierininae
Subfamily Chapmanininae
Family Calcarinidae
Family Elphidiidae
    Subfamily Elphidiinae
    Subfamily Faujasininae
Family Nummulitidae
    Subfamily Nummulitinae
    Subfamily Cycloclypeinae
Family Miogypsinidae
*Superfamily Orbitoidacea (restricted)
    *Family Lepidorbitoididae Vaughan, 1933 (nom. transl. Pokorný,
        1958, ex subfamily)
    Family Orbitoididae
    Family Discocyclinidae
    Family Lepidocyclinidae
        Subfamily Lepidocyclininae
        Subfamily Helicolepidininae
    Family Pseudorbitoididae
        Subfamily Pseudorbitoidinae
        *Subfamily Vaughanininae MacGillavry, 1963
*Superfamily Cassidulinacea (restricted)
    Family Pleurostomellidae
        Subfamily Pleurostomellinae
        Subfamily Wheelerellinae
    Family Annulopatellinidae
    Family Caucasinidae
        Subfamily Fursenkoininae
        Subfamily Caucasininae
    *Family Tremachoridae Lipps and Lipps, 1967
    Family Delosinidae
    Family Loxostomatidae
    Family Cassidulinidae
*Superfamily Nonionacea Schultze, 1854 (nom. transl. Subbotina,
    1959, ex subfamily)
    Family Nonionidae
        Subfamily Chilostomellinae
        Subfamily Nonioninae
    Family Alabaminidae
    Family Osangulariidae
    Family Anomalinidae
        Subfamily Anomalininae
        Subfamily Almaeninae

Superfamily Carterinacea
Family Carterinidae

VI. Acknowledgements

The writers wish to acknowledge our gratitude for research support by the Oceanography Section, National Science Foundation, NSF Grant GA-28756 (to Helen Tappan Loeblich). We thank Jeannie Martinez, Department of Geology, University of California Los Angeles for drafting the figure, and Carol Pincus for typing the manuscript.

VII. References

In the preceding discussion, many foraminiferal taxa are cited with author and date, not all of which are included in the references. Only articles and volumes to which other reference is made are included herein.

Anglada, R. and Randrianasolo, A. J. O. (1971). *Annls Univ. Provence, Sciences* **46**, 161–176.

Balakhmatova, V. T. (1972). *Vop. Mikropaleont.* **15**, 70–74.

Banner, F. T. (1966). *Vop. Mikropaleont.* **10**, 201–224.

Banner, F. T. (1970). *Revta. esp. Micropaleontologia* **2**, 243–290.

Bé, A. W. H. (1965). *Micropaleontology* **11**, 81–97.

Bé, A. W. H. and Ericson, D. B. (1963). *Ann. N.Y. Acad. Sci.* **109**, 65–81.

Bé, A. W. H. and Hemleben, C. (1970). *Neues Jb. Geol. Paläont. Abh.* **134**, 221–234.

Berchenko, O. I. (1971). *In* "Atlas fauny Turneyskikh otlozheniy Donetskogo Basseyna (s opisaniem novykh vidov)" (D. E. Ayzenberg, ed.) pp. 81–83. *Akad. Nauk ukr. RSR Inst. Geol. Nauk*, Naukova Dumka, Kiev.

Bermúdez, P. J. and Rivero, F. C. de (1963). "Estudio sistematico de los Foraminiferos Quitinosos, Microgranulares y Arenaceos." Edic. Biblio., Univ. Central Venezuela, Caracas.

Bignot, G. and Guyader, J. (1971). *In* Proc. II Planktonic Conference Roma (A. Farinacci, ed.), vol. 1, p. 79–83, Edizioni Tecnoscienza, Rome.

Blainville, H. M. D. de (1825). "Manuel de Malacologie et de Conchyliologie." F. G. Levrault, Paris.

Blow, W. H. (1969). *In* "Proc. First International Conference on Planktonic Microfossils" (P. Brönnimann and H. H. Renz, eds.), vol. 1, pp. 199–422, E. J. Brill, Leiden.

Brady, H. B. (1884). Repts. Scientific Results Voyage H.M.S. *Challenger*, vol. 9 (Zoology), pp. 1–814.

Brönnimann, P. and Zaninetti, L. (1971). *In* Brönnimann, P., Zaninetti, L., Bozorgnia, F., Dashti, G. R. and Moshtaghian, A. *Revue Micropaléont.* **14**, *Spec. No.* 5, 7–16.

Brotzen, F. (1963). *In* "Evolutionary Trends in Foraminifera" (G. H. R. Von Koenigswald, J. D. Emeis, W. L. Buning, and C. W. Wagner, eds.) pp. 66–78. Elsevier, Amsterdam.

Caudri, C. M. B. (1972). *Eclog. geol. Helv.* **65**, 211–219.

Cimerman, F. (1969). *Micropaleontology* **15**, 111–115.

Civrieux, J. M. S. de and Dessauvagie, T. F. J. (1965). *Maden Tetkik v Arama Enstit. Yayinl.* **124.**
Conkin, J. E., Conkin, B. M., Sawa, T. and Kern, J. M. (1970). *Micropaleontology* **16,** 399–406.
Corliss, J. O. (1972). *Am. Zool.* **12,** 739–753.
Cummings, R. H. (1955). *Micropaleontology* **1,** 221–238.
Cushman, J. A. (1925). *Smithson. misc. Coll.* **77 (4),** 1–77.
Cushman, J. A. (1948). "Foraminifera their classification and economic use." Harvard Univ. Press, Cambridge.
Douglas, R. (1964). *J. Protozool.* **11,** 484–486.
Douglas, R. (1969). *Micropaleontology* **15,** 151–209.
Douglas, R. and Sliter, W. V. (1965). *Tulane Stud. Geol.* **3,** 149–164.
Eimer, G. H. T. and Fickert, C. (1899). *Z. wiss. Zool.* **65,** 527–636.
Eldredge, N. and Gould, S. J. (1972). *In* "Models in Paleobiology" (T. J. M. Schopf, ed.), pp. 82–115. Freeman, Cooper and Co., San Francisco.
El-Naggar, Z. R. (1971). *In* Proc. II Planktonic Conference Roma (A. Farinacci, ed.), vol. 1, p. 421–476, Edizioni Tecnoscienza, Rome.
Fuchs, W. (1969). *Verh. geol. Bundesanst., Wien,* 1969 (2), 158–162.
Fuchs, W. (1970). *Verh. geol. Bundesanst., Wien,* 1970 **(1),** 66–145.
Fuchs, W. (1971). *Abh. geol. Bundesanst., Wien* **27,** 1–49.
Gerke, A. A. (1957). *Sbornik Statey po Paleontologii i Biostratigrafii* **3,** *Nauchnoissled. Inst. geol. Arktiki (NIIGA),* 31–52.
Glaessner, M. F. (1963). *In* "Evolutionary Trends in Foraminifera." (G. H. R. Von Koenigswald, J. D. Emeis, W. L. Buning, and C. W. Wagner, eds.) pp. 9–24. Elsevier, Amsterdam.
Gonzàles-Donoso, J. M. (1969). *Revue Micropaléont.* **12,** 3–8.
Grönhagen, D. and Luterbacher, H. (1966). *Eclog. geol. Helv.* **59,** 235–246.
Haake, F. W. (1971). *J. Foramin. Res.* **1,** 187–189.
Hamaoui, M. (1973). *Bull. Centre Rech. Pau, SNPA* **7,** 337–359.
Hamaoui, M. and Saint-Marc, P. (1970). *Bull. Centre Rech. Pau, SNPA* **4,** 257–352.
Hansen, H. J. (1968). *Meddr dansk geol. Foren.* **18,** 345–348.
Hansen, H. J. (1970a). *Meddr Grønland, Komm. Vidensk. Undersøgel. Grønland* **193 (2),** *Grønlands Geol. Undersøgelse* **93,** 1–132.
Hansen, H. J. (1970b). *Biol. Skr.* **17 (2),** 1–16.
Hansen, H. J. (1972). *Lethaia* **5,** 39–45.
Hansen, H. J. and Reiss, Z. (1971). *Bull. geol. Soc. Denmark* **20,** 329–346.
Hansen, H. J. and Reiss, Z. (1973). *Revta esp. Micropaleontologia* **4,** 169–179.
Hansen, H. J., Reiss, Z., and Schneidermann, N. (1969). *Revta esp. Micropaleontologia* **1,** 293–316.
Hay, W. W., Towe, K. M., and Wright, R. C. (1963). *Micropaleontology* **9,** 171–195.
Haynes, J. R. (1973). *Bull. Br. Mus. nat. Hist., Zoology, Suppl.* **4,** 1–245.
Hedley, R. H. (1963). *Micropaleontology* **9,** 433–441.
Hedley, R. H., Parry, D. M., and Wakefield, J. St. J. (1967). *Jl. R. microsc. Soc.* **87,** 445–456.
Hedley, R. H., Parry, D. M., and Wakefield, J. St. J. (1968). *J. nat. Hist.* **2,** 147–151.

Hinte, J. E. van (1963). *Jb. geol. Bundesanst., Wien, Sonderbd.* **8**, 1–147.

Hofker, J. (1969). *Stud. Fauna Curaçao* **31**, 1–158.

Hofker, J. (1970). *Publs Natuurhistorisch Genoot. Limburg* **20 (1–2)**, 1–98.

Hofker, J. (1971a). *Publs Natuurhistorisch Genoot. Limburg* **21 (1–3)**, 1–202.

Hofker, J. (1971b). *Revta esp. Micropaleontologia* **3**, 35–60.

Hofker, J. (1972). "Primitive agglutinated foraminifera." E. J. Brill, Leiden.

Hottinger, L. (1967). *Notes et Memoires Serv. Géol., Maroc,* **209**, 1–168.

Howe, H. V. (1936). *J. Paleont.* **10**, 415–416.

Hull, D. L. (1964). *Syst. Zool.* **13**, 1–11.

Jenkins, D. G. (1973). *Revta esp. Micropaleontologia* **5**, 133–146.

Kahler, F. and Kahler, G. (1966a). *Fossilium Catalogus I: Animalia.* Pars **111**, 1–254.

Kahler, F. and Kahler, G. (1966b). *Fossilium Catalogus I: Animalia.* Pars **112**, 255–538.

Kahler, F. and Kahler, G. (1966c). *Fossilium Catalogus I: Animalia.* Pars **113**, 539–870.

Kahler, F. and Kahler, G. (1967a). *Fossilium Catalogus I: Animalia.* Pars **114**, 871–974.

Kahler, F. and Kahler, G. (1967b). *Annln naturh. Mus. Wien* **71**, 107–115.

King, K. Jr. and Hare, P. E. (1972). *Science* **175**, 1461–1463.

Koehn-Zaninetti, L. (1969). *Jb. geol. Bundesanst., Wien, Sonderbd.* **14**, 1–155.

Lipps, J. H. (1964). *Tulane Stud. Geol.* **2**, 109–133.

Lipps, J. H. (1965). *Tulane Stud. Geol.* **3**, 117–147.

Lipps, J. H. (1966). *J. Paleont.* **40**, 1257–1274.

Lipps, J. H. and Lipps, K. L. (1967). *J. Paleont.* **41**, 496–499.

Loeblich, A. R. Jr. and Tappan, H. (1964a). "Treatise on invertebrate paleontology, pt. C, Protista 2, Sarcodina chiefly 'Thecamoebians' and Foraminiferida." Vols. 1, 2. *Geol. Soc. Amer.,* Univ. Kansas Press, Lawrence.

Loeblich, A. R. Jr. and Tappan, H. (1964b). *Bull geol. Soc. Amer.* **75**, 367–392.

Loeblich, A. R. Jr. and Tappan, H. (1964c). *J. geol. Soc. India* **5**, 5–40.

Lynts, G. W. and Pfister, R. M. (1967). *J. Protozool.* **14**, 387–399.

MacGillavry, H. J. (1963). In "Evolutionary Trends in Foraminifera." (G. H. R. Von Koenigswald, J. D. Emeis, W. L. Buning, and C. W. Wagner, eds.) pp. 139–196. Elsevier, Amsterdam.

McGowran, B. (1966a). *Contr. Cushman Fdn. foramin. Res.* **17**, 77–103.

McGowran, B. (1966b). *Micropaleontology* **12**, 477–488.

McGowran, B. (1968). *Micropaleontology* **14**, 179–198.

Mamet, B. L. (1970). *Can. J. Earth Sci.* **7**, 1164–1171.

Mamet, B., Mikhailoff, N. and Mortelmans, G. (1970). *Soc. belge Geol. Paleont. Hydrol., Mem. en 8°* **9**, 1–81.

Mamet, B. L. and Rudloff, B. (1972). *Revue Micropaléont.* **15**, 75–114.

Maslakova, N. I. (1964). *Vop. Mikropaleont.* **8**, 102–117.

Miklukho-Maklay, A. D. (1961). *Dokl. Akad. Nauk SSSR* **138**, 655–658.

Miklukho-Maklay, A. D. (1963). "Verkhniy Paleozoy Sredney Azii." Leningrad Univ.

Moullade, M. (1965). *C. r. hebd. Seanc. Acad. Sci., Paris* **260**, 4031–4034.

Neumann, M. (1967). "Manuel de micropaléontologie des foraminifères (Systématique-Stratigraphique) I." Gauthier-Villars, Paris.

Neumann, M. (1972). *Revue Micropaléont.* **15,** 163–189.
Norling, E. (1966). *Sver. geol. Unders. C* **613,** *Års.* **60 (8),** 1–24.
Norling, E. (1968). *Sver. geol. Unders. C* **623,** *Års.* **61 (8),** 1–75.
Norling, E. (1972). *Sver. geol. Unders. Afh. 14:0,* **47,** 1–120.
Nørvang, A. (1957). *Meddel. fra Dansk Geol. Foren.* **13 (5),** 1–135.
Oberhauser, R. (1964). *Verh. geol. Bundesanst., Wien,* 1964 **(1–3),** 196–210.
Parker, F. L. (1967). *Bull. Amer. Paleont.* **52,** 115–208.
Pessagno, E. A. Jr. (1964). *Micropaleontology* **10,** 217–230.
Pessagno, E. A. Jr. (1967). *Palaeontogr. Amer.* **5 (37),** 245–445.
Pessagno, E. A. Jr. and Miyano, K. (1968). *Micropaleontology* **14,** 38–50.
Poyarkov, B. V. (1965). *Dokl. Akad. Nauk SSSR* **163,** 728–730.
Putrya, F. S. (1963). *Paleont. Zh.* 1963 **(1),** 35–41.
Putrya, F. S. (1970). *Paleont. Zh.* 1970 **(4),** 29–45.
Raup, D. M. and Stanley, S. M. (1971). "Principles of Paleontology." W. H. Freeman and Co., San Francisco.
Rauzer-Chernousova, D. M. and Fursenko, A. V. (1959). "Osnovy Paleontologii. Obshchaya chast' Prosteyshie." Akad. Nauk SSSR, Moscow.
Redmond, C. D. (1964). *Micropaleontology* **10,** 251–263.
Reiss, Z. (1958). *Micropaleontology* **4,** 51–70.
Reiss, Z. (1963). *Bull. geol. Survey Israel* **35,** 1–111.
Reiss, Z. and Luz, B. (1970). *Revta esp. Micropaleontologia* **2,** 85–96.
Rozovskaya, S. E. (1969). *Paleont. Zh.* 1969 **(3),** 34–44.
Seiglie, G. A. (1964a). *Carib. J. Sci.* **4,** 497–512.
Seiglie, G. A. (1964b). *Boln Soc. geol. Mexicana* **27,** 1–9.
Seiglie, G. A. and Bermúdez, P. J. (1965a). *Boln Inst. Oceanogr. Univ. Oriente* **4,** 155–171.
Seiglie, G. A. and Bermúdez, P. J. (1965b). *Geos* **12,** 15–65.
Sliter, W. V. (1968). *Paleont. Contr. Univ. Kans. ser.* **49,** *Protozoa* **7,** 1–141.
Soliman, H. A. (1972). *Revue Micropaléont.* **15,** 35–44.
Srinivasan, M. S. (1966). *Trans. R. Soc. N. Z. Geol.* **3,** 231–256.
Steineck, P. L. (1971). *Texas J. Sci.* **23,** 167–178.
Subbotina, N. N. (1971). *In* "Novoe v sistematike mikrofauny." *Trudȳ vses. Neft. nauchno-issled. geol.-razv. Inst.* **291,** 63–69.
Tappan, H. and Lipps, J. H. (1966). *Bull. Am. Ass. Petrol. Geol.* **50,** 637.
Tappan, H. and Loeblich, A. R. Jr. (1966). *Vop. Mikropaleont.* **10,** 375–392.
Tappan, H. and Loeblich, A. R. Jr. (1968). *J. Paleont.* **42,** 1378–1394.
Towe, K. M. (1971). *In* Proc. II Planktonic Conference Roma (A. Farinacci, ed.), Vol. 2, pp. 1213–1224, Edizioni Tecnoscienza, Rome.
Towe, K. M. and Cifelli, R. (1967). *J. Paleont.* **41,** 742–762.
Voloshina, A. M. (1970). *Paleont. Zh.* 1970 **(4),** 108–110.
Vyalov, O. S. (1966). *L'vov Univ. Paleont. Sbornik* **3 (1),** 3–11.
Vyalov, O. S. (1968). *Dopov. Akad. Nauk ukr. RSR* 1968 **(1),** ser. *B,* 3–6.

Biometry of the Foraminiferal Shell

G. H. SCOTT

New Zealand Geological Survey Lower Hutt. N.Z.

I. Is Biometry Relevant?	55
II. Chamber Morphology	58
A. Proloculus in Smaller Foraminifera	58
B. Embryonic Chambers in Larger Foraminifera	60
C. Nepionic Chambers in Larger Foraminifera	62
D. Chamber Wall	67
E. Chamber Dimensions	73
F. Apertural Architecture	75
G. Retrospect on Chamber Studies	75
III. Shell Geometry	76
A. Turbinate Shells of Some Smaller Foraminifera	76
B. Plane Spiral Shells	93
C. Has Measurement Supplanted Words and Pictures?	99
IV. Data and Analysis	100
A. Collection	100
B. Operator Error	102
C. Univariate Methods	103
D. Shape and Size	109
E. Mapping Populations: Distance and Discrimination	119
F. Classification	124
V. Population Studies in Taxonomy and Stratigraphy	128
A. Taxonomy and Biometry: Theory and Practice	128
B. Biometry, Stratigraphy and Evolution	133
C. Shell Design	139
D. Testing the Synthetic Theory	144
VI. Acknowledgments	144
VII. References	144

I. Is Biometry Relevant?

In the history of research on the foraminiferal shell, biometry has played a minor, often insignificant role. Intensive study of the taxonomy of the order, especially of fossil representatives, over the last fifty years has resulted in unparalleled progress in global classification of upper Paleozoic, Cretaceous and Tertiary strata. Yet the taxonomic foundations were not built with biometrical techniques. For example, none of the taxa employed in the Tertiary zonations discussed by Blow (1969) and Berggren (1971) were recognized biometrically although certain aspects of the variability of some species (e.g. coiling direction) were quantified. Although many quantitative data have been collected for Paleozoic fusuline taxa they have not been analysed and applied in the classificatory process. In the main, foraminiferal taxa have been recognized intuitively from qualitative assessment of shell characters and applied in biostratigraphy without quantitative criteria for population variability. To the extent that the utility of these stratigraphical classifications justifies the taxonomic procedure, the first, historical, answer to the question "is biometry relevant?" must be negative.

This answer will satisfy only the prejudiced seeking confirmation of the view that the qualitative techniques long espoused by foraminiferal taxonomists do not require rethinking. It does not explain why the vast majority of students have not found biometry relevant to their work nor does it justify similar neglect in the future. For this, the past record of biometrical work must be examined. Foraminiferal taxonomy is based almost exclusively on characters of the shell. Biometry, considered broadly, deals with quantitative analysis of biological data. A first step is to consider those aspects of the foraminiferal shell that have been quantified. Do they sufficiently characterize the shell for taxonomic studies? Can the "gestalt" approach of the intuitive taxonomist be matched by a set of measurements? Are there problems in measurement of characters and collection of data? Of no lesser importance are the analytical techniques employed. Have the data so far collected been adequately analysed? Are the peculiar requirements of the taxonomist met by techniques currently in use? Are there analytical models available that have been neglected by foraminiferal biometricians? Can future developments in collection and analysis be identified?

Answers to these questions will illuminate the present status of biometry at the technical level, but there are philosophical considerations that suggest that biometry should occupy a much more central role in foraminiferal research than hitherto. In taxonomy much hinges on the unit of classification. Is it the population or the individual? The theses of the "new systematics" and the synthetic theory are built around the population. Population variability is taxonomic information. The approach should be multivariate and statistical. Subjective assessment of variation is inadequate, in principle and practice, for scientific communication. Here then, is a major role for biometry. The revolution in methodology in mid-twentieth century initiated by Huxley (1940) produced resonances from like-minded foraminiferal taxonomists (Glaessner, 1955; Emiliani, 1969) but little expansion in biometrical research. Were there constraints at the technical level or was the philosophy rejected by the majority of workers? Certainly, in much of foraminiferal taxonomy, the holotype still occupies greater classificatory significance than any other specimen. I suggest that the relevance of biometry to taxonomic practice depends, in part, on acceptance by students of the view that objective studies of population variability are an important component of research.

The role of biometry in research on the foraminiferal shell is also determined by the questions posed. Most research on fossil taxa has been directed at classification of strata. Biometrical studies have followed suit. Thus trends in morphology of lineages have been mapped for stratigraphical information. To the contrary, the history of lineages, often excellently preserved, has neither been investigated for insights into evolutionary processes and patterns nor for explanations of shell architecture. These intrinsically biological topics have been neglected and immediate stratigraphical objectives preferred. Yet, in perspective, stratigraphical classification by fossils is an applied field which is likely to benefit by results from these biological studies. Does the form of the foraminiferal shell reflect the adaptive response of the organism to physiological and external environmental constraints? Why are there changes in shape during ontogeny of some shells? Are they size-required? Investigation of these and allied questions calls for quantitative analysis of shell form, a blend of statistical studies of ontogenetic and phylogenetic patterns, deterministic modelling of shell architecture and evaluation of the efficiency of structure for known or hypothesized functions. The methodology has been admirably expounded by Gould (1970).

Closely similar shell shapes have evolved repeatedly (Cifelli, 1969; Drooger, 1956). In the Cenozoic history of the Foraminiferida at least, major novelty in architecture was exceptional. So prevalent were parallel and convergent trends that, at the present stage of comprehension, supraspecific classifications are effectively horizontal (grouping of similar architectural forms from different lineages) rather than vertical (grouping of taxa according to evolutionary history). Unravelling of parallel and convergent trends will require multivariate mappings of lineages rather than the univariate delineations previously applied in stratigraphical research. In conjunction with quantitative analyses of form and function, these mappings may be expected to contribute to detailed understanding of foraminiferal evolution, a subject as yet grossly neglected. Is geographic isolation an important component in speciation? Do morphological trends in lineages represent improvements in adaptation? Is there evidence in support of non-selectionist or orthogenetic factors orienting the direction of evolutionary trends? Is ontogeny a key to phylogeny? Such topics are important in framing of evolutionary theory and require quantitative investigation.

It will be to the detriment of foraminiferal research if quantitative methodology is eschewed in the future as it has been in the past. Thus I wish to identify the problems currently impeding the application of biometry as well as to suggest its future role. Biometry may have been irrelevant in the past. This status will be raised as research becomes more concerned with explanatory descriptions of shell form and variation.

II. Chamber Morphology

Multichambered (polythalamous) taxa predominate amongst the foraminifera. In these taxa the shell is built by accretion of chambers, often of similar architecture, formed at successive instars. At death or reproduction a complete record of skeletal development is available for study. Biometrical analysis of the shell can thus be made at two levels: the morphology of the structures forming the chamber, the basic unit of shell architecture, and the organization of the chamber sequence to form the shell.

A. Proloculus in Smaller Foraminifera

In the few cases analysed, foraminifera have displayed relatively complex life histories (Hedley, 1964 summarized the literature) with sexual and asexual phases. In some taxa, specimens arising from sexual reproduction may exhibit shell features distinct from those of asexual origin. The size of the proloculus (the initial chamber formed in ontogeny) is one such character. Hofker (for bibliography, see Hofker, 1968) worked extensively on the problem of recognition of polymorphism from data on the proloculus. Measurements on the greatest diameter of the proloculus were plotted as size frequency distributions and as bivariate scatters against gross shell dimensions. Generally, the size frequency distributions contained several modes. Intermodal minima were applied as boundaries between generation types. Dimorphic and trimorphic schemes were developed from the data. The first methodological problem concerns delimitation of the taxon sampled for polymorphism. Hofker's procedure was to recognize a taxon qualitatively and then sample proloculus size. A change in the taxonomic concept is likely to affect the shape of the distribution of proloculus size. The problem is recondite for fossil material where there is no experimental evidence about the life cycle. In an extreme example of polymorphism in a lagenid taxon, quoted by Hofker (1968), each form would, in conventional horizontal classification, be placed in a separate genus. However, granted the veracity of the population concept employed, the technical problem in Hofker's analyses concerns recognition of clusters from univariate and bivariate data. Visual inspection has been employed. This is adequate for certain data. For other data (Fig. 1) intermodal minima as selected by Hofker seem ill-defined and an objective clustering procedure is required to test the validity of the partitions. The lack of such tests probably led Freudenthal (1969, p. 58) to complain that "whatever the merit of the theory on trimorphism in foraminifera may be, it must be borne in mind

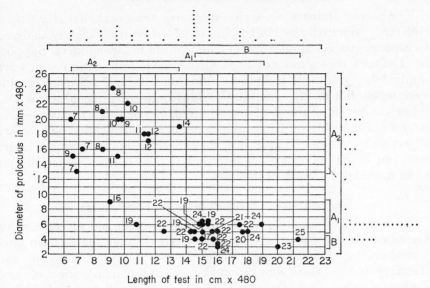

Fig. 1. Length of shell plotted against diameter of proloculus in *Bolivina robusta* (bottom samples, apparently from several tropical localities). Weak separation of megalospheric forms A_1, A_2. (After Hofker, 1951).

that convincing statistical evidence has never been presented on the basis of morphology, neither by Hofker, nor by any other author". Such rejection may be unwarranted. There are isolated data on prolocular dimensions (Reyment, 1959 on *Afrobolivina afra*) that are trimodal. But a caveat is necessary: the frequency distribution (Fig. 2) pools material from many localities and does not establish that 3 modes occur in a given population.

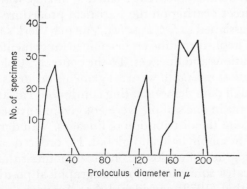

Fig. 2. Distribution of proloculus diameters in *Afrobolivina afra* (pooled localities). Mode at left represents microspheric form, modes at centre and right interpreted as 2 megalospheric groups. (After Reyment, 1959).

The present status of research is that large amounts of data have been collected, principally by Hofker, but have not been sufficiently analysed to support general application of trimorphism in foraminifera. Support for Hofker's theses has not been forthcoming from others who have applied his graphical techniques (Adams, 1957). Additional, multi-variate, studies on the interrelationships between proloculus size and other characters of the shell are required. Recognition that some taxa may be dimorphic or trimorphic is vital in classification. There are objective techniques (chapter IV) for cluster recognition but the larger problem is that of biological interpretation of clusters, especially for fossil material, rather than statistical.

B. Embryonic Chambers in Larger Foraminifera

In orbitoidal foraminifera the initial protoconch and the succeeding chamber (deuteroconch) are usually clearly distinguishable from later chambers by their size and shape. Several Upper Cretaceous and Tertiary lineages show convergent changes in the arrangement of these embryonic chambers (Drooger, 1956). In *Lepidocyclina*, van der Vlerk (1959; 1963; 1968; van der Vlerk and Postuma, 1967) estimated the extent to which the deuteroconch encloses the protoconch (Fig. 3). Van der Vlerk and Bannink (1969) applied similar measurements to *Oper-culina*. The various measurements proposed to represent amount of enclosure (embryonic acceleration of Drooger, 1956) are simple but require accurate sections (Drooger and Freudenthal, 1964). Assessment of their relative merits for mapping evolutionary trends has yet to be made (Freudenthal, 1972). Similarly, operator error in preparation and measurement has not received the attention it warrants.

As a correlative technique, advocated by van der Vlerk, grade of enclosure is subject to a caveat concerning adequacy of stratigraphic control in initial delineation of the trend and linkage with stage classifications. But of direct bearing on the statistical procedure are the limits of variation to be assigned to *Lepidocyclina*. Van der Vlerk's schemes separate eulepidine and nephrolepidine embryonic arrangements as representing separate populations and lineages. To the contrary, Cole (1960) claimed that in *Lepidocyclina radiata* these arrangements occur as intrapopulation variants, although data demonstrating continuous variation between the two arrangements in one sample were not given. Furthermore, there are divergent views on the distribution of lineages. Van der Vlerk regarded *Lepidocyclina* as monophyletic and global. He ordered strata throughout the world using grade of enclosure on the hypothesis that equal grade of enclosure indicates equivalence in stratigraphical position. To the con-trary, Freudenthal (1972) considered that American and Euro-African populations evolved independently and therefore that equality of enclosure did not necessarily imply stratigraphical equivalence. Para-

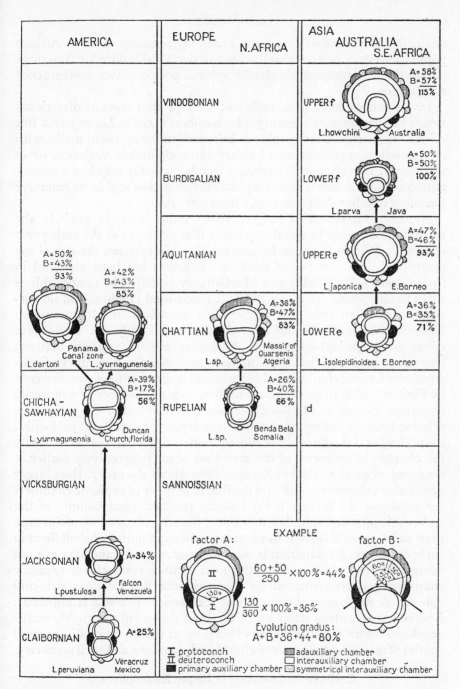

Fig. 3. Embryonic chambers in *Lepidocyclina*: a method for measuring extent to which the deuteroconch encloses the protoconch and regional history of increment in enclosure. (From van der Vlerk, 1959).

doxically, Freudenthal's data for grade of enclosure of West African populations does integrate well with van der Vlerk's data for American populations of comparable stratigraphical position. Was convergence synchronous?

Investigations have been sufficient to show that marked directional trends in arrangement of embryonic chambers occur in *Lepidocyclina*. But efforts to extract stratigraphical information have been made with inadequate comprehension of either intrapopulation variation or of phylogeny. Detection of convergent trends probably requires a combination of more detailed stratigraphical investigation and more extensive sampling of other shell characters than hitherto.

Nepionic chambers (Drooger, 1956) either directly encircle the embryonic or occur in spiral sequences that originate at the embryonic chambers. Within lineages, inverse relationship between the size of the protoconch and number of nepionic chambers has been observed in *Cycloclypeus* by Tan (1932) and MacGillavry (1956). The relationship is weak in *Miogypsina* (Drooger, 1963). Freudenthal (1969) found no correlation between reduction of nepionic chambers in Mediterranean Neogene *Planorbulinella* and size of embryonic chambers. These data relate to megalospheric forms. A similar inverse relation between increment in size of the megalospheric proloculus and number of subsequently formed chambers with a particular arrangement was observed by Hofker (1968) in smaller foraminifera (e.g. the number of multiserial chambers formed in *Marssonella oxycona*). Research into the volumetric relation between embryonic chambers in larger foraminifera (proloculus in smaller) and the immediate subsequently spiral chambers may clarify the changes in geometry of the shell that occur progressively earlier in ontogeny of quite unrelated lineages throughout the order. Does larger prolocular volume coupled with decline in number of nepionic chambers (or analogues in smaller taxa) lead to constant total volume of the embryonic plus nepionic chamber suite? Structural questions of this type need to be posed if there is to be progress in explanation of shell form in early ontogeny. Explanation is inhibited because relevant data are not being collected. As we see in the next section, research on nepionic chambers has concentrated on their number not on their size and volume. A fundamental criticism of biometrical practice is apparent. Research has centred on facets of the shell that seem to provide stratigraphical information. This has often been at the expense of integrated studies of relationships between chamber dimensions and shell geometry.

C. Nepionic Chambers in Larger Foraminifera

Tan (1932) made a classic investigation into the number of nepionic chambers in Indonesian populations of megalospheric *Cycloclypeus*. The work arose from a report by van der Vlerk that the number of nepionic chambers was of classificatory value. Numbers of nepionic septa in samples were tabulated or plotted as frequency polygons (Fig. 4). The data showed several important features: (i) frequency distributions were polymodal; (ii) the same modes recurred in successive samples in a

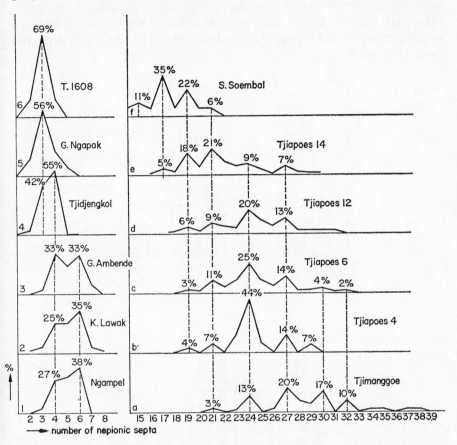

* The graph no. 4 (Tjidjengkol) refers to the population of *Cycl. postindopacificus* var. *tenuitesta*. In graph b (Tjiapóes 4) the maximum indicated by 7% is erroneously drawn at the *abscissa* 29. It should be removed to the *abscissa* 30.

Fig. 4. Polymodal frequency distributions for nepionic septae in *Cycloclypeus*, samples arranged stratigraphically. Note coincidence of modes, sample to sample, and progressive shift in range. (After Tan, 1932).

stratigraphical sequence (undescribed); (iii) there was an overall stratigraphical trend towards decrease in number of septa. Tan took consistency in location of modes among samples to indicate the coexistence of a number of mutants or "elementary species". A saltational theory of evolution was invoked to explain their origin and their ranges were applied to detailed classification of strata.

Cosijn (1938) applied Tan's approach to Spanish material. His data confirmed (i) and (iii) above but not (ii), although several coincident modes appeared in frequency distributions for area of the protoconch in section. The relation between distribution of nepionic septa and external morphology of the shell in one sample led Cosijn to conclude that two taxa were present. But it must be noted that distributions for the two species, as separated on these other features of the shell, still departed from unimodality.

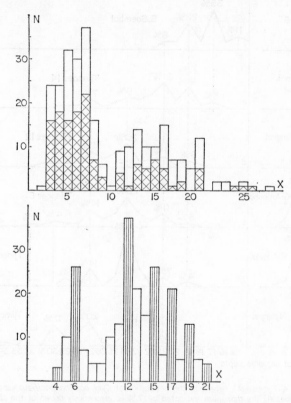

Fig. 5. (Upper) *Cycloclypeus eidae* group, Neogene, Borneo; histogram of number of nepionic chambers. Cross-hatched segments refer to specimens in which counts were most reliable. (Lower) Data for same group from Tan (1932). Modes for Tan's elementary species marked by vertical rulings. (From Drooger, 1955).

Drooger (1955) examined a sample of the *Cycloclypeus eidae* group from the same province as Tan's material. Replicated counts of nepionic septa were performed. Only those specimens providing consistent data were used in construction of frequency histograms. Drooger obtained a bimodal frequency distribution for numbers of nepionic chambers and for diameter of protoconch (Fig. 5). From comparison with Tan's data for the *Cycloclypeus eidae* group, Drooger suggested that both distributions were primarily bimodal, with the intermodal minimum in the same location. Within each mode, class frequencies were much more irregular in Tan's than in Drooger's data. Drooger agreed with Cosijn that the local maxima in Tan's data (Tan's "elementary species") may have been due to operator bias. The major modes were interpreted as arising from presence of microspheric and megalospheric generations (the number of nepionic chambers is inversely related to the size of the protoconch).

MacGillavry (1962) reanalysed data collected by Tan, Cosijn and Drooger in terms of evolutionary lineages identified from a plot of mean diameter of protoconch (logarithmic data) against mean number of nepionic chambers. Several lineages were identified with Tan's "elementary species" and intralineage progression, at varying rates, towards reduction of nepionic chambers was found. Some lineages, according to MacGillavry, arose saltatively. Drooger's data were reinterpreted to indicate the coexistence of two lineages, not dimorphism.

The outstanding problems that emerged in the history of biometrical investigation of *Cycloclypeus* are those of operator variability in measurement, resolution of polymodal frequency distributions and the biological interpretation to be placed on groups recognized biometrically. The work also raises questions about the loss of biological information that may result from restriction of studies to the earliest stages of ontogeny of the shell and about the emphasis of enumerative features of these early chambers to the exclusion of consideration of their shape and growth features. The resolution of the problems raised by Tan's work offer some of the most interesting investigations in foraminiferal biometry. Nevertheless his major thesis, that within many lineages of larger foraminifera (Tan, 1939) there is progressive reduction in number of nepionic chambers, is now widely accepted as a principle.

Nepionic acceleration has been investigated in greater detail by Drooger (e.g. 1952; 1963 with bibliography) from global studies of mid-Tertiary miogypsinids. Souaya (1961), Ujiié and Oshima (1960) and Ujiié (1966) made additional biometrical studies. Lineages were recognized by Drooger, in part following Tan (1936), on the presence or absence of lateral chambers and the position of embryonic chambers relative to the periphery of the shell. Within lineages, biometrical investigation concentrated on the number of nepionic chambers (where a single protoconchal spiral was present) and on the symmetry of spirals where two arose from the protoconch. Other enumerative and angular relations of nepionic spirals and embryonic chambers were investigated but none provided comparable discrimination of populations. Biometry of the shell and features of its later growth have been neglected. Indeed there seems to have been *a priori* reasoning in support of recapitulation. Drooger (1952, p. 8) wrote that "it is reasonable to suppose that these early ontogenetic features will better show hereditary characters of the organisms than all others, occurring in later ontogenetic stages". External features of the shell were considered to be unreliable because they may have been influenced by environmental conditions. It is true, as Drooger noted, that concentration on nepionic features eliminates a source of variation due to inclusion in the sample of specimens at different stages of growth. His samples include only specimens in which nepionic growth has been completed. But it is yet to be shown biometrically that

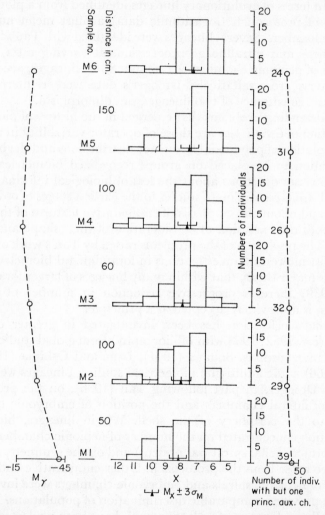

Fig. 6. Histograms show number of nepionic chambers in *Miogypsina tani* from successive samples at Moulin d'Eglise, France. There is a gradual, progressive decline in mean number but little change in range. (After Drooger *et al.*, 1955).

classificatory information resides primarily in the early stages of growth. Nor has it been explained why variation in features of later growth seemingly lacks a genetic basis.

Drooger (1963) considered that the Miogypsinidae were monophyletic with some local branching from the main stock of *Miogypsina* s.s. Parallel trends in nepionic acceleration, at differential rates, were found in the local branches. In the main stock, acceleration in populations from widely separated basins may have been comparable.

Nepionic trends in American populations (Drooger, 1952) were determined with little information about the stratigraphical order of localities. More satisfactory evidence was found in samples (Fig. 6) from successive strata in southwestern France (Drooger *et al.*, 1955). But even within this region, *prima facie* evidence of discordant nepionic data from populations in strata correlated on other evidence emerged. The sporadic distribution of miogypsinids, possibility of reworked populations and inadequate independent stratigraphical control inhibits reconstruction of the phylogeny of the group and stringent testing of nepionic acceleration. These problems also diminish the value of *Miogypsina* in global correlation. Nevertheless, there are indications that valuable interregional correlations may arise from Drooger's biometrical studies. Stratigraphical relationships indicated by New Zealand lower Miocene populations (Drooger, 1963; Freudenthal, 1969) accord with data from planktonic taxa. No other benthic group appears to provide such interregional resolution of mid-Tertiary strata.

Cole (1964, p. 138) complained that Drooger's biometrical classification "is inflexible, as the species are recognized by mathematic averages which do not reflect the variation which may occur in the basic growth pattern between individuals of a single population". He recognized three species in American mid-Tertiary rocks whereas Drooger recognized fifteen. Cole's subsequent discussion reveals a different approach to the classification of nepionic chambers but fails to show how Drooger's statistical measures do not reflect variation in numbers of nepionic chambers. As Ujiié (1966) remarked, a reasonable criticism of nepionic acceleration initially requires the application of Drooger's methods. This Cole failed to do.

D. Chamber Wall

1. COMPOSITION

Gross mineralogical composition of secreted carbonate chamber walls (Blackmon and Todd, 1959) offers little scope for biometrical analysis as variation occurs only at high taxonomic levels. Quantitative analyses of amino acid composition of planktonic shells (King and Hare, 1972) revealed marked interspecific variation which was attributed to changes in genotype. Intraspecific variation was not assessed but this biochemical approach may provide discriminatory taxonomic information at much lower levels than has appeared from mineralogical or element analyses.

Analysis of variability in type, size and orientation of particles cemented into walls of agglutinating taxa offers much scope but has not advanced. Buchanan (1960) found that in species of *Schizammina* and *Jullienella*, particle size in walls matched closely with that of the sediments on which they lived. Whether such random incorporation of particles is typical of foraminifera is unknown.

2. STRUCTURE

Structure of the chamber wall has become of first order importance in classification of calcareous taxa above the specific level (Loeblich and Tappan, 1964). Towe and Cifelli (1967), in a general theory of wall ultrastructure, suggested that the optical dichotomy radial: granular employed by Loeblich and Tappan does not necessarily imply a morphological dichotomy in structural type and arrangement. They showed that the optically granular wall has a preferred orientation with calcification taking place on a rhombohedral face while the optically radial wall exhibits preferred orientation with calcification taking place on a basal face. Hansen (1968) collected X-ray diffractometer data that agreed with Towe and Cifelli's model. Preferred orientation in radial walls may be variable but data adequate for analysis of variability are not yet available.

Lynts and Pfister (1967) distinguished between an outer pavement layer and an inner randomly oriented layer in the porcellaneous wall of certain miliolids. Bivariate statistics on the length and width of calcite rhombs composing the pavement layer showed that their shape differed between taxa.

3. THICKNESS

Ujiié (1963) recorded that the thickness of the walls of the ultimate and penultimate chambers in Neogene planktonic taxa was closely similar but found no significant relation between diameter of the shell and thickness of the walls. He did not distinguish lamellae deposited at formation of the penultimate chamber from those due to construction of the ultimate chamber. Orr (1969) found that thickness of the penultimate chamber wall of *Globigerinoides ruber* (bottom samples, Gulf of Mexico) increased by 8–10 μ across the continental shelf (Fig. 7). The material was standardized for size. He considered that the data reflected increments due to secondary calcification as the water column deepened. In ontogenetic series, average wall thickness tended to increase rapidly with size.

There is an absence of data on the relation between interior dimensions of chambers and gross wall thickness; on variations in thickness of a lamella over the exterior walls of the shell and on the relation between wall thickness and shape of the chamber. These and similar data are essential for systematic study of shell mechanics.

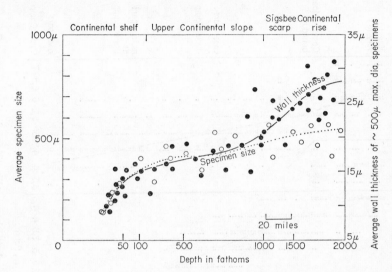

Fig. 7. Increase in average shell size and in wall thickness of penultimate chamber in *Globigerinoides ruber*, from seaward traverses in Gulf of Mexico (bottom samples). (After Orr, 1969).

4. PORES

Pores are tubular openings penetrating the chamber wall; their functions are uncertain and may vary among higher taxa. Angell (1967a) considered that the pore processes in *Rosalina floridana* welded the wall elements into a single structure; Hofker (1968) attributed to pores a respiratory function.

Little numerical data has been collected on shape of pores. An approximately circular cross-section is typical but Lynts and Pfister (1967) gave a bivariate characterization for pores in *Fursenkoina punctata* that showed that average shape was elliptical with the major axis twice the length of the minor.

Progressive increase in pore diameters of Cretaceous gavelinellid taxa was noted by Hofker (1957; 1968). The pattern of increase was gradual during the history of a species (Fig. 8). Some descendant taxa, at their divergence, reverted back to pores with small diameters and then showed the same gradual increment (Fig. 9). Hofker attributed this pattern to greater respiratory requirements as each species increased in size.

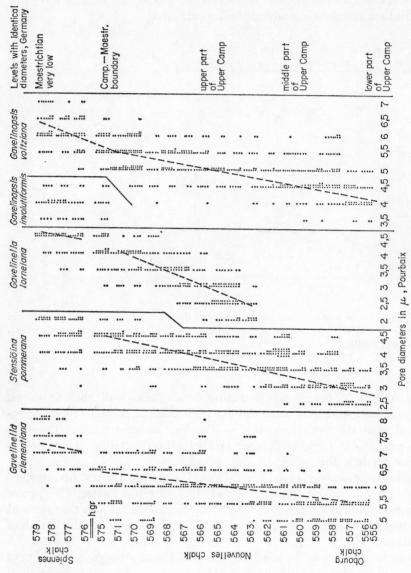

Fig. 8. Progressive increment in pore diameters, shown by gavelinellid species in an upper Cretaceous section, Pourbaix, Belgium. (After Hofker, 1968).

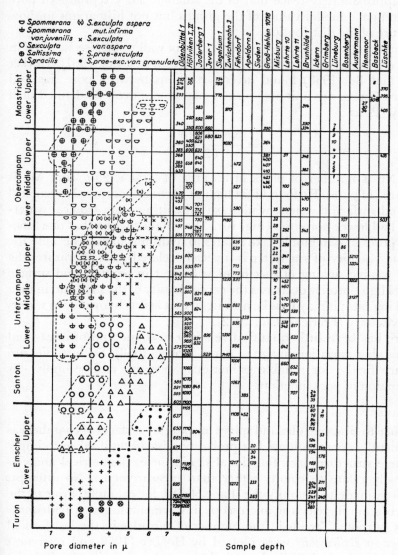

Fig. 9. *Stensiöina* from Cretaceous sections in western Europe: increments in pore diameters in successive species. Hofker considered divergences occurred when a pore diameter of about 4 μm was attained in ancestral taxa. (After Hofker, 1957).

Whereas Hofker's data for pore dimensions are unimodal, Reyment (1959) and Berggren (1960) have suggested from studies of *Bolivina* and *Globigerina* that two size groups are present, the linear dimension of the larger approximately twice that of the smaller. They reproduced pore fields and gave modal values of the groups but not histograms to show the nature of the distribution.

The concentration of pores per unit area of wall in *Globigerina eggeri* in Pleistocene cores varies closely with temperature (Wiles, 1967). Hofker (1968) and Bé (1968) found inverse relations between pore size and number of pores per unit area among planktonic taxa. Bé defined shell porosity (P) as

$$P = \frac{Np\ Ma\ 100}{625}$$

where Np is number of pores in an area of $25\ \mu \times 25\ \mu$, and Ma is mean area of pores. He found that highest porosities occurred in tropical taxa, and surface dwellers, while taxa inhabiting extra tropical water-masses had lower porosities (Fig. 10). These data suggest that one function of pores in planktonic taxa relates to hydromechanics. Because the dynamic viscosity of sea water is inversely related to temperature, the rate of passive sinking of specimens is highest in tropical and surface waters. High porosity decreases the excess density of the shell and in turn the sinking rate; it may thus be a strategy that assists buoyancy in waters of low viscosity (Frerichs *et al.*, 1972).

There has been little quantitative study of shape and of intra- and interchamber variation in size of pores (Zobel, 1968). Such studies are necessary firstly, for reliable estimates of porosity and secondly, for investigation of pore function.

5. ORNAMENT

Reference to ribs, pustules and spines on exterior surfaces of chambers as "ornament" is conventional but probably misleading if taken to imply that these structures lack functional significance. Biometrical analyses principally relate to counts rather than measurements. No attempts have been made to quantify the shape of ribs, for example. Prior to the advent of scanning electron microscopy, instrumentation was inadequate for this purpose. Trends in the number of pustules in the last-formed chamber in *Bolivinoides* were plotted by Hofker (1968 and references therein). Kroboth (1966) showed that numbers of ribs on *Citharina* progressively increased within stratigraphical sections. Hofmann (1971) analysed a reticulate rib pattern in *Bolivina* by measuring gross length of the pattern on the shell and maximal inter rib spacing. Lutze (1964) measured lengths of spines and costae on *Bolivina argentea*. The material for these studies was not standardized for stage in ontogeny. Consequently, little was revealed about ontogenetic variation. Because the purpose of the studies was related to taxonomy and empirical charting of lineages, the design of the structures in relation to chamber and shell mechanics was neither considered quantitatively nor analysed. As with

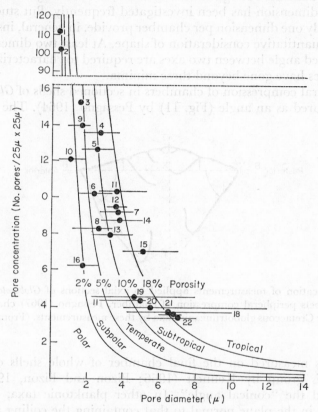

Fig. 10. Pore concentration plotted against diameter for 22 living planktonic taxa. Horizontal lines indicate observed ranges of pore diameters. (After Bé, 1968).

most chamber and shell characters investigated biometrically, empirical classificatory information has been sought at the expense of explanatory interpretations of form.

E. Chamber Dimensions

In polythalamous taxa, chambers constitute the basic building unit of the shell. As the shape of the building unit directly influences the architecture of the shell, study of chamber shape is important. Yet little quantitative research has been accomplished. In larger foraminifera the field is virgin. Discussing measurable features of nummulitid shells,

Drooger *et al.* (1971, p. 47) wrote, "The shape of chambers and septa appeared extremely difficult to express numerically. Several methods were tried, but they appeared to give very inexact results, to be rather time consuming, and too hard to employ by many people". The relation between a dimension of a given chamber and that of an adjacent chamber or a shell dimension has been investigated frequently. But studies that involve only one dimension per chamber provide, in general, insufficient data for quantitative consideration of shape. At least two dimensions or the included angle between two axes are required to characterize shape. Few studies have gone beyond these minima.

Peripheral compression of chambers in sectioned shells of *Globorotalia* was measured as an angle (Fig. 11) by Pessagno (1964). The opposite

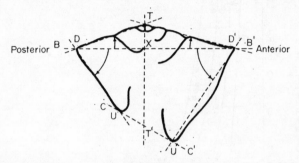

Fig. 11. Location of measurements applied to axial sections of *Globorotalia*. Angle XD'U' reflects peripheral compression of chamber. Pessagno (1967) characterized many upper Cretaceous globotruncanid taxa by these measurements. (From Pessagno, 1964).

angle was measured on the final chamber of whole shells of the *G. crassaformis* group by Kennett (1966; Bizon and Bizon, 1971) and designated the "conical angle". In other planktonic taxa, chamber diameters in the plane normal to that containing the coiling axis were investigated by Gorbatschik (1964) and Scott (1972a). Scott also collected longitudinal diameters (parallel to the coiling axis) of the last-formed chamber. Globigerinid taxa commonly adopt particularly simple chamber shapes that can be approximated by truncated ellipsoids. Four variates (one specifying truncation) should be sufficient for such chambers. But generally chamber shape will not be approximated by a simple geometric figure and the number of data required to reconstitute form may be large. There has been a complete lack of research in this field. Technical problems apparent are orientation of free specimens or sections and the location of homologous reference points for measurements. The latter problem is particularly acute and a solution may lie in use of scanning techniques wherein measurements are taken at selected intervals around the wall with reference to the chamber centroid.

F. Apertural Architecture

Although the aperture is commonly the largest sub-structure of the chamber it is commonly at least one magnitude smaller in maximum dimension. With conventional low magnification stereomicroscopy, data may be subject to considerable error because the maximum opening of the aperture occupies only about one division on the measurement graticule. In multiserial turbinate taxa the aperture is often located at the inner margin of the chamber. Oblique viewing may be required and difficulties in achieving consistent orientation from specimen to specimen arise. There are a few data for globigerinid taxa in which the aperture is relatively large. Berggren and Kurtén (1961) showed that there was positive correlation between the width of the aperture in the final chamber and diameter of the shell in *Globigerina yeguaensis*. Scott (1970) found that height and width of the primary aperture (final chamber) in *Globigerinoides trilobus* were positively correlated with the diameter of this chamber. Large changes in height of the primary aperture occurred in this species (Scott, 1968, 1970), and the area of the opening, relative to gross chamber dimensions, may have decreased during evolution.

The meagre data available on apertural dimensions throw no light on the following important topics: (i) relation between size of the opening and size either of the chamber in which it occurs, or of the succeeding chamber (the aperture is a major access-way for protoplasm during chamber morphogenesis); (ii) ontogenetic changes in shape of the aperture and their relation to changes in chamber shape.

G. Retrospect on Chamber Studies

The greatest impact on foraminiferal research arising from biometrical study of chambers has been Hofker's study of prolocular size in relation to polymorphism. Although his particular theory of trimorphism is disputed by most workers, in part because it outruns inadequately analysed data, he has established that polymorphism is a problem that requires quantitative data for its solution. This is important at a stage in foraminiferal research when measurement is usually ignored.

Whereas Hofker has researched a general problem, other biometrical studies of chambers have related primarily to empirical taxonomy or to delineation of morphological trends for stratigraphical purposes. Only Tan's work on nepionic chambers has produced a concept of wide application.

The failures are obvious. Characterization of shape has been neglected. There is no indication yet that biometry can replace and improve on the eye of the practised taxonomist. Certainly in this respect biometry is still irrelevant. In turn, because shape has not been quantified neither has it been analysed. Large, thin-walled chambers might well be viewed, mechanically, as thin shells. There has been no analysis of the mechanical properties of particular shapes, the extent to which loads are sustained by membrane stresses and of methods of reinforcement. These studies may offer insights on the functional significance of shape.

III. Shell Geometry

A. Turbinate Shells of Some Smaller Foraminifera

Following Thompson (1952) I refer to shells in which the spiral formed by the chamber sequence translates along the coiling axis during growth as turbinate. These are frequently called trochoidal in conchology but

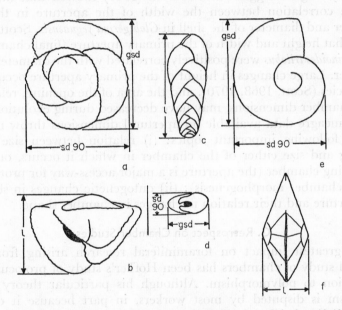

Fig. 12. Terminology for gross dimensions of shells of smaller foraminifera. a b, *Globorotalia* (turbinate multiserial); c, d, *Bolivina* (turbinate biserial); e, f, *Lenticulina* (planispiral—after Marie (1941) who discussed geometry). gsd, greatest spiral diameter; sd 90, spiral diameter at 90° to gsd; l, length.

this term also refers to a curve generated by a point in the plane of one curve that rolls on another. Terminology for measurements on turbinate shells has varied according to numbers of chambers per revolution (uni-, bi-, tri-, multiserial). To maintain uniformity in this review I follow these definitions (Fig. 12). Length (1) is the maximum dimension of the shell, parallel to the coiling axis. Greatest spiral diameter (gsd) is the greatest diameter in the plane normal to that containing the coiling axis. Usually, gsd intersects the last-formed chamber. It is commonly called "width" or "breadth" in biserial shells. Spiral diameter at a right angle to the greatest (sd90): this diameter is measured in the same plane as gsd. Often, it has been termed "thickness".

1. BISERIAL SHELLS

Smith (1963) examined large suites of recent or subfossil Bolivinidae from the Pacific Ocean and used methods of analysis that were later applied by Lindenberg (1966) and Hofmann (1967) to Cenozoic species. The data collected on the family are large and the methods typical of much biometrical research on foraminifera. The Bolivinidae thus provide a good example from which to assess the biometrical contribution to the taxonomy of smaller foraminifera and the understanding of shell form.

Smith collected data on four variates: shell length, greatest spiral diameter, number of chambers and prolocular diameter. Other measurements were considered but found difficult to replicate. This is a common difficulty in foraminiferal biometry and reflects problems in orientation of the shell for measurement and in defining homologous loci. The small number of variates prejudices adequate quantitative characterization of shell form.

Statistics on length were computed in detail (Fig. 13) but produced little significant taxonomic information. This is expectable in view of the very wide age segment (as judged by number of chambers) present. For some species, differences in mean length among stations could be due to variations in age structure. Statistics on length for specimens with a given number of chambers may have been much more informative. However, plots of length against numbers of chambers provided good discrimination between megalospheric and microspheric forms in some taxa. Shape of the shell was characterized by length:width ratios and by regression lines for these variates. These techniques discriminated between flaring shells (little change in length:width ratio throughout size range) and parallel-sided shells (ratio increases with shell length). The most important taxonomic result of this study was the finding that, for some taxa, shape varied between populations from various depths. Similar results were obtained by Lutze (1964) and interpreted as a response to geographic variation in oxygen concentration. Some basinal populations exhibited retarded shell (less flare) growth in comparison with populations inhabiting slope biotopes.

It is significant, methodologically, that the biometrical analyses led to no major changes in classification. Why? The material was initially identified using extant qualitative diagnoses and quantitative data were used primarily to supplement the existing taxonomy. This is an essential methodological point, applicable to most work in foraminiferal biometry. Biometry has not been applied to the construction of taxonomies,

Sample and depth	I. Abundance			N	IIa. Length statistics (mm)					IIb. Length frequency M = mean length	III. Number of chambers	IV. Width	V. Flare
	Total in split	Perfect in split	Total in sample		Range	M	MD	σ	V				
C-15; 435 meters	43	22	172	57	S.R. 0.15–0.48 O.R. 0.16–0.62	0.31 ±0.01	0.04	0.06 ±0.01	17.5 ±1.6				
C-8; 450 meters	123	110	3936	110	S.R. 0.12–0.55 O.R. 0.18–0.50	0.33 ±0.01	0.06	0.07 ±0.004	21.3 ±1.4				
C-14; 800 meters	7	6	448	32	S.R. 0.10–0.42 O.R. 0.15–0.38	0.26 ±0.01	0.04	0.05 ±0.01	20.2 ±2.5				
C-10; 1700 meters	33	25	33	25	S.R. 0.09–0.48 O.R. 0.18–0.45	0.29 ±0.01	0.05	0.08 ±0.01	23.5 ±3.3				

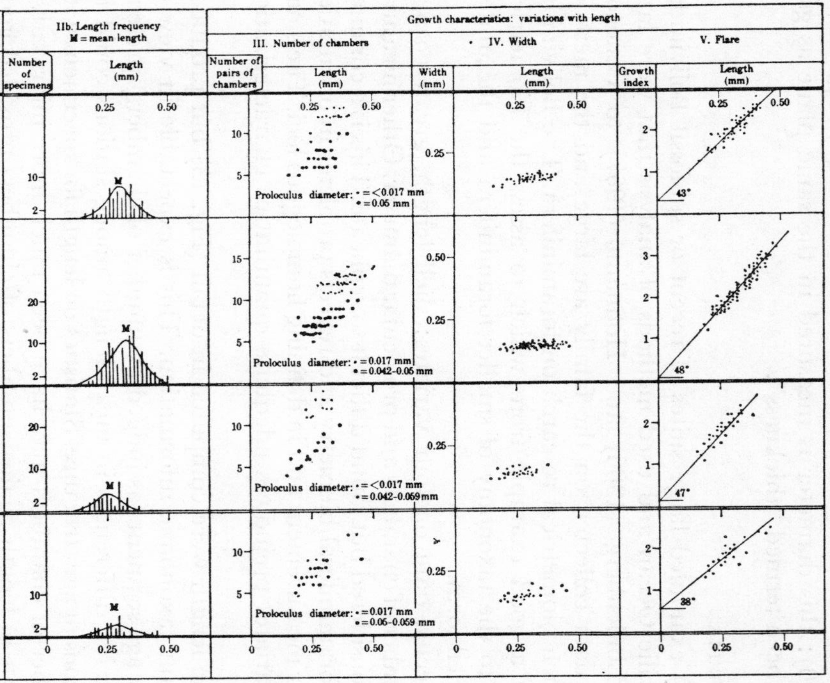

Fig. 13. *Bolivina humilis*: length statistics, relationships between length, width (greatest spiral diameter) and the ratio of length and width (growth index) in bottom samples from various depths. In this species, width shows little variation with length. (From Smith, 1963).

rather existing qualitative taxonomies have been accepted as the point of departure, whether for supplementation, or for testing. It is exceptional to find the biometrician attempting to classify *de novo*.

On the basis of qualitative taxonomy, Smith reported the occurrence of at least twelve bolivinid species in six square inches of bottom sediment and considered that their coexistence indicated response to microenvironments. The real biometrical challenge with such data is surely to produce a quantitative classification of the bolivinid population as a basis for taxonomic interpretation. I do not suggest that there would necessarily be a 1:1 relation between groups recognized biometrically and taxa of species rank. Some might be, for example, intrapopulational clusters reflecting size or age.

Lindenberg and Hofmann improved description of the shell by including data on width (sd90) but the biometrical characterization remained supplementary to qualitative "words and pictures" (Smith, 1963, p. A11). Note also that all three authors measured the shell at conclusion of growth and that a wide size spectrum was represented in most samples. The quantitative data on shell shape (e.g. length:width ratios and regressions) thus relates to shape at conclusion of growth of shells of various sizes. The history of shape in shells at various stages of their growth was not investigated. This neglect is again typical of biometrical research on smaller foraminifera. In turn, it has inhibited interpretative studies of the adaptive significance of shell shape.

Little is known of changes in form, during ontogeny, in other biserial taxa. The ratio between gsd:sd90 in a Miocene lineage of *Textularia* (Kennett, 1963) was found to change unidirectionally in time. However, length and number of chambers were not considered so that the behaviour of the ratio during growth of individuals is unknown.

2. TRISERIAL TO UNISERIAL SHELLS

Meulenkamp (1969) studied in detail the change in number of chambers formed per revolution during growth of the shell in Mediterranean Neogene lineages of *Uvigerina*. Well defined changes, during ontogeny, were found from triserial to irregular uniserial, to regular uniserial arrangements (Fig. 14). The relative length of the shell occupied by uniserial chambers showed a trend to increase during the history of the lineages. The relative number of primitive uniserial chambers tended to decrease in higher populations but there was no sustained trend for length of shell to increase. As the arc between successive chambers was not measured, it is uncertain whether the trends reflect a change in angle from 120° (triserial) directly to 360° (uniserial) or whether intermediate angles occurred. It is possible that the angle between primitive uniserial chambers is less than 360°. Although spiral geometry is not known in detail, the data indicate that uniserial arrangement occurred progressively earlier in the growth of the shell.

Frequency polygons of the number of biserial chambers were used by Grabert (1959) to trace the connection between ancestral *Gaudryina* (triserial with a few biserial chambers formed late in growth) and

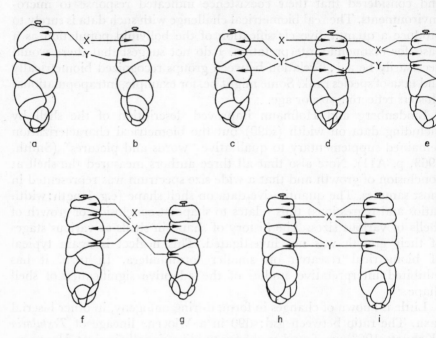

Fig. 14. Change of chamber arrangement in late ontogeny of *Uvigerina* from triseria to uniserial. y: primitive uniserial chambers; x: fully developed uniserial chambers (From Meulenkamp, 1969).

Spiroplectinata (triserial stage greatly reduced or absent, and long biserial, occasionally uniserial stages). The method nicely summarizes gross phylogeny (Fig. 15) but disregards concomitant changes in chamber and shell shape. Again, the partial nature of the biometrical data is shown.

3. BISERIAL TO UNISERIAL SHELLS

For a textularid lineage Scott (1965) used the ratio between the diameter of the last chamber and gsd to measure the extent to which biseria chamber arrangement had moved towards uniserial arrangement. A gradual, unidirectional trend towards uniserial arrangement was shown but the data collected were quite inadequate for an understanding of the geometry underlying the change.

Reduction in number of chambers per revolution has occurred commonly, and independently, in the phylogenies of tri-, bi-, or multiseria

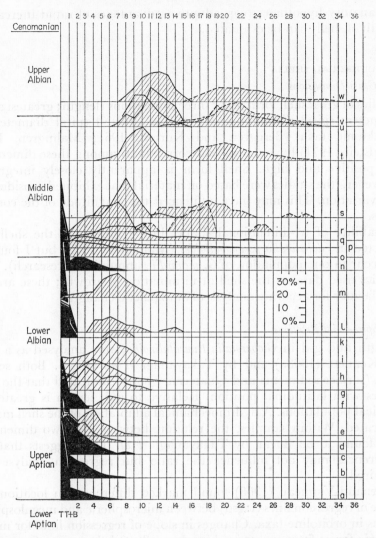

Fig. 15. Progressive increase in number of biserial chambers (1, 2, 3, . . .) in phylogeny of *Gaudryina-Spiroplectinata* lineages. Black shading: ancestral populations of *G. dividens* and *S. lata*. Wide hachures: *G. compacta* (geographically isolated taxon). Narrow hachures: progressive populations of *G. dividens* and *Spiroplectinata spp.* (After Grabert, 1959).

shells in which the rate of translation along the coiling axis was already high. Students have concentrated on describing the progress of the reduction in successive populations, usually for biostratigraphical purposes. The mechanics of the reduction have in no instance been analysed in relation to shell geometry and chamber shape. This is necessary before explanatory hypotheses (e.g. adaptive advantages of rapid increase in length of shell) can be evaluated.

4. MULTISERIAL SHELLS

(a) Gross Dimensions

Studies of gross shell form using three dimensions (length; greatest spiral diameter; spiral diameter at 90° to the greatest spiral diameter) in planktonic taxa (Berggren and Kurtén, 1961; Malmgren, 1972; Gradstein, 1971) have shown that correlations among these dimensions are positive and high. Shell form is apparently closely integrated. However, the studies were based on material that exhibited considerable size variation. This may have influenced the magnitude of the correlations.

Stability of tri-dimensional shape during growth of the shell was rejected in *Globorotalia pseudobulloides* (Malmgren, 1972) but I found it to occur in one population of *G. menardii* (unpublished research). Both studies used data on the shell at termination of growth; these are unsatisfactory for accurate analysis of ontogenetic changes.

(b) Spiro-axial Relations

Length and gsd in *Globorotalia truncatulinoides* were analysed as a ratio by Kennett (1968a) and by Takayanagi *et al.* (1968). Both sets of data (surface sediments and Quaternary cores) indicated that the ratio varies with latitudinal position. Relative to gsd, length is greatest in tropical populations. A hydromechanical adaptation of the shell may be indicated. Within samples, the relation between the two dimensions was found to be linear by Takayanagi *et al.* This suggests that size differences may not influence the ratio but regression analyses are required.

Arnaud-Vanneau (1968) found marked differences in location and slope of regression lines (l:gsd) between microspheric and megalospheric forms in orbitoline taxa. Changes in slope of regression lines for microspheric forms from successive horizons reflected the onset of growth of annular chambers.

Two components of shell length (Fig. 16) were defined for *Praeglobotruncana* by Klaus (1960). The ventral component (h' in Fig. 16) is determined by the angle made by the chamber profiles with the coiling axis. This variate may be subject to considerable operator variation when chamber profiles are inflated. Polymodal plots of the ratio between dorsal length and gsd provided partial evidence of the contemporaneous presence of several taxa with overlapping variation fields. Similar results from samples of *Rotalipora* were obtained by Caron (1967) in a companion study.

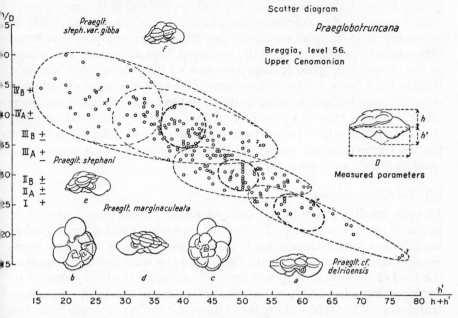

Fig. 16. Shell measurements in *Praeglobotruncana*. Continuous variation field exhibited by several taxa from one sample although the distribution of h/D is polymodal. (After Klaus, 1960).

(c) *Spiral Diameters*

In populations of *Globigerina eocaena*, Lindenberg (1969) found that correlation between gsd:sd90 was influenced by variation in rate of expansion of chambers. Bivariate scatters for equivalent measures on *Globorotalia miozea miozea* and related taxa (Scott, 1972a) showed the same influence: specimens in which the final chamber was smaller than the penultimate were offset from the main scatter. It appears that these diameters are highly correlated when the ratio between given dimensions (*d*, say diameters) of consecutive chambers ($i - 1, i, i + 1, \ldots n$) is constant throughout the sample for at least the final $270°$ of each spiral. Variation in the ratio d_n/d_{n-1}, particularly, produces concomitant variation in spiral diameters of shells. These observations indicate the intimate relation between shell form and behaviour of the chamber sequence. The former is the integral of the latter.

The final chamber is commonly smaller than the penultimate in specimens of *Globigerapsis index* and there may be a continuous gradation between small final chambers and bullae. Quilty (1969) analysed the relation between gsd and sd90 both in "normal" individuals (final chamber greater in size than penultimate) and those with a variety of final chambers (or bullae) smaller than the penultimate. In the latter groups he ignored the final chamber and applied gsd to the penultimate chamber and made a corresponding change in location of sd90. Study of regression lines showed no significant differences in slope, but some differences in position ($P < 0.01$) occurred. The data show that shape at the penultimate growth stage as estimated by gsd and sd90 is closely similar for a wide range of specimens in which the final chamber is diminutive or bulla-like. Modifications to the final chamber have been cited as generic criteria in the *Globigerapsis* group. Quilty's data suggest that individuals which would have been assigned to different genera at the final growth stage, would, at the penultimate growth stage, have been assigned to the same species.

(d) The Chamber Sequence

In most smaller foraminifera, each chamber added during growth is larger than its predecessor. The rate of expansion can be studied using homologous measurements made on chambers formed at known positions in the sequence. Michael (1966) used the ratio between length of the final chamber and length of the shell excluding the final chamber to describe shape in species of *Gavelinella*. These shells exhibit very slight translation of the spiral along the coiling axis so that the ratio is almost wholly a measure of rate of expansion of chamber length over the last 180° of the spiral.

Research on the chamber sequence, particularly in late ontogeny, in planktonic taxa is the most detailed. Berggren and Kurtén (1961) found that while the diameters of the final (n^{th}) and $(n-2)^{th}$ chambers (*Globigerina yeguaensis*) were positively correlated, with product moment correlation coefficients significantly different from zero at the 0.01 level, weaker correlations, probably not significantly different from zero, were obtained for the $n^{th}:(n-3)^{th}$ chambers. When the diameter of the $(n-2)^{th}$ chamber was held constant, negative coefficients, sometimes significantly different from zero, were obtained for the relation between $n^{th}:(n-3)^{th}$ chambers. Variations in relative size of the nth chamber were recognized as causative. Comparable results for the same variates were obtained by Scott (1970) from *Globigerinoides trilobus* and by Malmgren (1972) from *Globorotalia pseudobulloides*. Adjacent chambers showed significant positive correlations (0.01 level) but as the interval between chambers increased, coefficients decreased in value.

The distribution of $\log p = \frac{1}{2} \log (d_n/d_{n-2})$, where d_n is the diameter of the nth chamber and d_{n-2} that of the $(n-2)^{th}$ chamber, in a sequence of populations of *Globigerinoides trilobus* and orbulinids was examined by Scott (1966b). Crudely unimodal frequency distributions included higher values as the stratigraphic sequence was ascended but very low values (<0.02) persisted. Note that the transformation used does not make variance independent of the mean and its merit requires investigation. Values of mean $\log p$ increased with stratigraphical position as did variance. In the lower *G. trilobus* populations differences between the actual diameter of the $(n-1)^{\text{th}}$ chamber and the theoretical, d'_{n-1} $(=p(d_{n-2}))$ were small and random. This indicates that $d_n/d_{n-1} \simeq d_{n-1}/d_{n-2}$. Much greater departures occurred in higher samples and suggested that the ratios were no longer stable. The presence of orbulinids in these samples is significant. A partial explanation is that in orbulinids, d_n/d_{n-1} is typically much greater than d_{n-1}/d_{n-2}. Large negative deviations of d'_{n-1} from d_{n-1} are expectable in such specimens. Lindenberg (1969) similarly analysed d_{n-3} and d_{n-1} in *Globigerina eocaena* and found small, and probably random, departures between theoretical and expected values of d_{n-2}. However, analysis of d_n and $d'_n (d'_n = p(d_{n-1}))$ in the same material revealed strongly negative departures. This was due to frequent occurrence of specimens in which d_n/d_{n-1} was less than d_{n-1}/d_{n-2}.

Plots of chamber dimension against chamber position provide a type of growth curve that elucidates patterns more clearly than the foregoing studies of ratios between pairs of chambers. In *Globigerinoides* and *Globorotalia* examined by Scott (1970, 1972a), curves (on arithmetic co-ordinates) which are regularly concave upward, without inflexions, are rare. Typical are curves that have segments of increasing slope followed by segments of declining slope. It is common for the slope between the $n-1)^{\text{th}}$ and n^{th} chambers to be less than that between the $(n-2)^{\text{th}}$ and $(n-1)^{\text{th}}$ chambers $(d_n/d_{n-1} < d_{n-1}/d_{n-2})$. Note, however, that the $(n-1)^{\text{th}}$ chamber is not the sole locus at which inflexion occurs. Olsson (1971; 1972) collected data on the diameter of chambers in *Globorotalia fohsi* over the complete chamber sequence. Logarithmic diameters of early chambers, up to about the 10th, plotted linearly against their angular position in the sequence. This indicates that increments in log diameter are constant during early growth. Later chambers also form a linear trace but it is displaced, and in some specimens of steeper slope than that for early chambers (Fig. 17). Displacement may denote alteration in chamber form and change of slope indicate introduction of an increment rate differing from that employed in early growth. The size of the final chamber was frequently smaller than the penultimate.

In summary, from analysis of a few planktonic taxa it appears that, excluding the final, chambers form an expanding sequence with $d_{i+1} \simeq k d_i$ as Thompson (1952) maintained. Growth curves, if k is constant, are linear on semilogarithmic coordinates (exponential form on arithmetic coordinates). On this model, correlation between dimensions of successive chambers of stable shape should be high. The evidence for decline in correlation as the interval between chambers increases requires further examination. It may reflect variation in k or it may be due to "noise" created by inclusion in samples of specimens not standard-ized for age, or even numbers of chambers. Comparison of $(n - 4)^{th}$ and n^{th} chambers in a sample where n is not defined, and perhaps highly variable specimen to specimen, may lead to spurious estimates of cor-relation. Full ontogenetic data, such as those collected by Olsson, are much more satisfactory than the "mixed" data used by other workers.

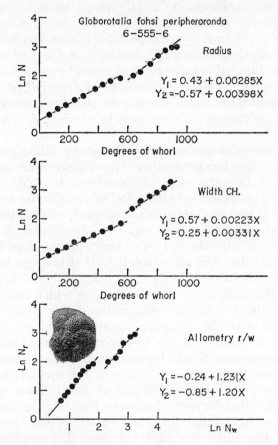

Fig. 17. Radius of shell (upper) and width of chamber (centre) plotted for each chamber added during growth of a specimen of *Globorotalia fohsi peripheroronda*. The relative growth plot (lower) indicates positive allometry of radius relative to chamber width and suggests that there is a major alteration in chamber shape at the tenth chamber. (After Olsson, 1972).

The last formed chamber in planktonic taxa is exceptional in that $d_n/d_{n-1} < d_{n-1}/d_{n-2}$ occurs commonly. Berger (1971) found a maximum of "kummerforms" (specimens in which $d_n/d_{n-1} < 1$) off California in a layer of northern water sandwiched between southern water masses. He suggested that kummerforms were related to environmental stress in the northern mass. However, there are no data on the hydrographic location of such specimens at the time at which the small final chamber was constructed. Contrary to Berger, Olsson (1973) considered that the kummerform condition signified attainment of adult size. This view does not account for occasional occurrences of $d_i < d_{i-1}$, prior to $i = n$, in some ontogenies. Greater understanding of the causes of low (for the individual), values of d_n/d_{n-1} is important as the presence of a small final chamber has been employed as a taxonomic criterion (Jenkins, 1967a). Whereas the small final chamber may be without adaptive advantage to the organism, it is likely that the mid-ontogenetic change in increments of diameter in *Globorotalia fohsi* reflects some size-correlated improvement in shell mechanics. Disentangling of phenotypic and genotypic effects on the chamber sequence has yet to commence.

(e) *Number of Chambers Per Revolution*
Increments of wall material at each instar over the external surface of the shell of many turbinate species tend to render early chambers

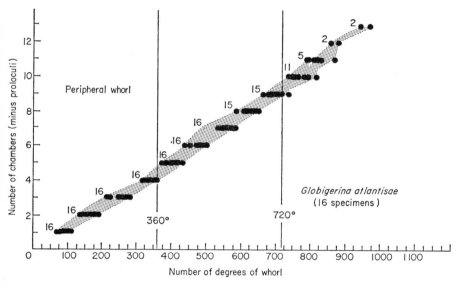

Fig. 18. Angular location of chambers in *Globigerina atlantisae*. (After Cifelli and Smith, 1970).

increasingly obscure as ontogeny progresses. The most reliable data thus relate to later ontogeny. Cifelli and Smith (1970) graphed the angular location of chambers backward from the last-formed. For *Globigerina atlantisae* the plot (Fig. 18) is approximately linear whereas for *G. incompta* it exhibits upward concavity due to decrease in number of chambers in the last revolution. Cifelli and Smith considered that this reduction characterized most globigerinids. However, note that the number of chambers in the final revolution of some compressed species of *Globorotalia* (e.g. *G. multicamerata*) may be higher than in preceding revolutions. The changes in geometry of the spiral that produce increments or decrements in number of chambers per revolution have not been analysed. Alterations in d_i/d_{i-1}, relocation of the centre of the chamber generating curve relative to coiling axis and changes in chamber shape are some possible mechanisms.

Number of chambers in the final revolution is frequently cited in taxonomic descriptions. The following example is cautionary. Lewis and Jenkins (1969) found that in modern populations of *Nonionellina* in the New Zealand region, the number of chambers in the final revolution gradually increased with latitude. There was a corresponding increase in size of the shell. Sectioned specimens from a southern population showed ontogenetic increase in number of chambers per revolution. The contrast in number of chambers in the final revolution between northern and southern populations was due to the extended ontogeny, as judged by total number of chambers, of southern specimens in a species that exhibited an ontogenetic trend towards increase in number of chambers per revolution. Latitudinal variation in number of chambers in the final revolution of shells of *Globigerina pachyderma* was described by Kennett (1968b; Malmgren and Kennett, 1972), but connection with earlier growth of the shell was not established. Olsson's data (1971) for this species do not show conspicuous ontogenetic change in number of chambers per revolution.

(f) Number of Chambers and Shell Dimensions

Because they are added consecutively to the shell, the number of chambers is a function of the age of the organism. The total number built may be species-specific but there are few enumerative data to establish this. Certainly, populations in culture have shown wide variability in number of chambers built prior to reproduction (Bradshaw, 1957). Frequency distributions of number of chambers in shells from dead assemblages may reflect selective preservation, sampling and differential mortality during growth, as well as any modal number at which reproduction is most likely to occur. Hansen (1970) showed that the distributions of chamber numbers in fossilized shells of *Globoconusa daubjergensis* were usually unimodal. The corresponding distributions for greatest spiral diameter of the shell were grossly unimodal but displayed "sawtooth" irregularities. More analyses are required to explicate the relation between chamber number and gross shell dimension. Proloculus size must be considered as it forms the origin of the chamber sequence. Expectably, chamber number and gross shell size should be very highly correlated in specimens in which chamber shape and rate of expansion are constant throughout ontogeny.

Size of the shell at termination of growth may be related to environmental factors (Lewis and Jenkins, 1969). Nicol (1944) recognized subspecies of *Elphidium* partly on maximum size of the shell but Burma (1948) reinterpreted the data as an example of clinal variation related to water temperature. In data for planktonic taxa (Orr, 1967; 1969), modal size of the shell in dead populations from recent sediments showed a trend to increase with distance from shore and with depth. The trend reflects, in part, decreased representation of small shells but the relation between shell size and chamber number remains to be investigated. Small shells appear to be under-represented in sediments relative to their frequency in the plankton (Berger, 1971). This may indicate that mortality is low in early ontogeny; alternatively (or additionally), it may reflect selective preservation of larger shells.

Size variation of another form was described by Eicher (1960). Marked variation in greatest spiral diameter of *Trochammina depressa* for specimens with fixed chamber number occurred among samples from several fossil biofacies. Smallest diameters occurred in a biofacies, probably of lowered salinity, with a depleted number of species. Such data may imply interpopulation variation in shell parameters such as d_i/d_{i-1}. With Eicher, it may be appropriate to regard as dwarfed, populations that for a given chamber number, are smaller in some major shell dimension than is typical for the taxon (or generation thereof if it is polymorphic).

Hofker (1968, and references therein) stressed that phyletic increase in size of the shell occurs commonly in foraminifera. For example, the diameter of the last chamber in *Bolivinoides* from a Chalk section showed gradual translocation in range and increment in mean value (Fig. 19). Hofker linked increment in shell size with that of the megalospheric proloculus. But more elaborate data than say, a single shell dimension, are needed to analyse the mechanics of the trend. Two possible strategies, that could be used either singly or in concert, are increment in rate of chamber expansion (d_i/d_{i-1}) and extension of the chamber sequence.

(g) Coiling Direction

Study of the direction of coiling in turbinate shells is the simplest, and probably the most widely applied, biometrical technique in foraminiferal research. Cosijn (1938) observed marked fluctuations in the direction of coiling in *Globorotalia menardii* from Neogene strata and subrecent sediments in Indonesia but the value of data on coiling directions for studies of planktonic taxa was established by Bolli (1950; 1951; 1971) who described the principal trends and plotted data for many taxa

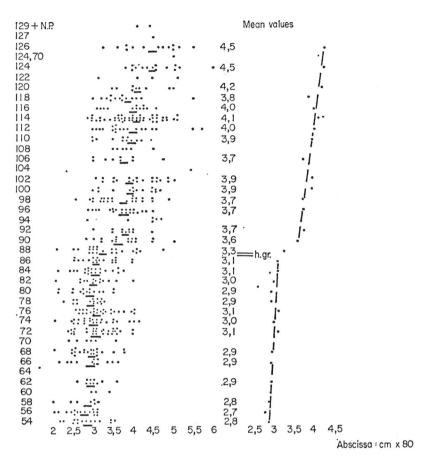

Fig. 19. Gradual increment in diameter of last chamber in a lineage of *Bolivinoides* from a Cretaceous section at Glons, Belgium. (After Hofker, 1959).

(Fig. 20). Applications have been to local stratigraphic correlation (Nagappa, 1957; Bandy, 1960; Jenkins, 1967a; 1967b, and paleoclimatic interpretation (Ericson *et al.*, 1963).

Raw data (% sinistral or dextral) are commonly plotted against stratigraphical or geographical location. Kennett (1968b) used pie graphs. There has been little statistical analysis. Simple tests of significance using binomial probability paper (Mosteller and Tukey, 1949) for departures of sample data from hypothetical values (e.g. random coiling) could be usefully applied, especially when sample sizes are small. Ujiié (1963) calculated correlation coefficients between numbers of sinistral specimens: total number of specimens of that species for planktonic

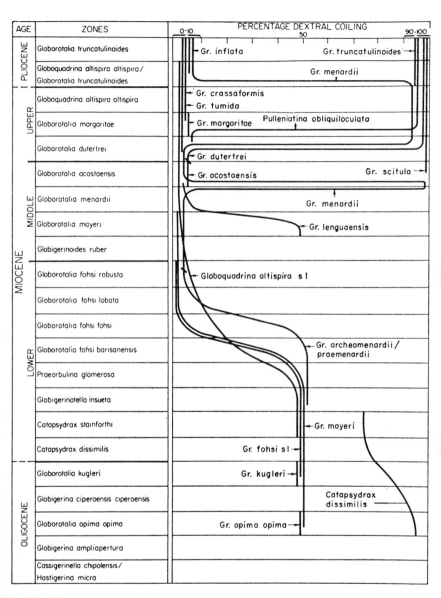

Fig. 20. History of coiling for some species of *Globorotalia*, *Globoquadrina*, *Catapsydrax* and *Pulleniatina*. Random coiling early in the history of some taxa tends to be replaced later by strongly preferred directions. (From Bolli, 1971).

species. For most taxa, correlations were highly significant. He inferred that this indicated constant environmental conditions in the strata sampled. Of several characters investigated in *Globorotalia pseudobulloides* (Malmgren, 1972) only length of the shell showed significant correlation with direction of coiling. I do not know of a multivariate discriminatory assessment of the magnitude and significance of differences in shell morphology between dextral and sinistral specimens in the same population. Also lacking is any study of the effect of selection on coiling direction. Does coiling direction in juvenile populations equate with their adult counterparts? This is significant, practically, in elucidating the effect of age distribution in the sample on coiling ratio. It may also contribute to understanding of marked geographic fluctuations in the ratio noted in populations of living species.

Coiling direction appears unsuitable as a primary stratigraphical criterion when the species on which it is to be determined shows repeated oscillations in direction. Jenkins (1967a) established a sequence of subzones using repeated oscillations in the direction of coiling of *Globorotalia pachyderma*. A set of subzones (Dj) characterized by dextrally coiled populations alternate with subzones (Sj) characterized by sinistrally coiled populations to give a sequence D1, S1, D2, . . ., D5. The subzones are of little value in stratigraphy unless the full sequence is present in a section or supplementary criteria are invoked. Biostratigraphical correlation depends on non-repeating criteria. An isolated sample with dextrally coiled specimens could represent any of the D1–5 subzones. Also of concern to the stratigrapher are suggestions that coiling direction is linked to environmental parameters. If so, then unless these are constant over the entire geographic distribution of a taxon, false correlations will be made. Bandy (1968) proposed that a temperature model governs the pattern of coiling in *G. pachyderma* and Kennett (1968b) viewed changes in coiling direction in this species as a cline related to latitude. To the contrary, Cifelli (1971) maintained that there is no fixed relationship between coiling direction in *G. pachyderma* and temperature. Kaever (1960) found weak relation between coiling direction and environment in Cretaceous *Globorotalites*. Correlation by fossils usually requires criteria that are non-repeating in time and stable geographically. The first condition is satisfied by the coiling history of some taxa but for most taxa the second is unsubstantiated.

As yet, biometrical effort has been primarily directed toward mapping trends in coiling directions. There seem to be few technical problems in this work provided that populations can be objectively defined and that coiling direction is not influenced by the age distribution of specimens. It seems that little progress has been made in interpretation either in relation to modern environmental parameters or phylogenetically. For example, Hofker (1972) disputed Bolli's thesis that random direction of coiling prevails early in the history of planktonic taxa. Finally, it is curious that so much attention has been paid to coiling direction in planktonic shells and so little to the behaviour in benthic shells.

B. Plane Spiral Shells

1. FUSULINES

The shells of many upper Paleozoic fusuline taxa form a plane spiral. Conventionally, taxonomic research has been based on sections of the shell, either in a plane containing the coiling axis (axial section) or

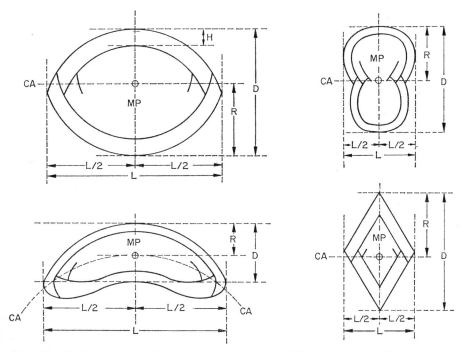

Fig. 21. Variates commonly measured in sections of fusuline shells: diameter, D; radius, R; chamber height, H; length, L; half length L/2. MP is median plane, CA is coiling axis. (After Cutbill and Forbes, 1967).

normal thereto (sagittal or spiral section). Well located sections display the history of shell growth volution by volution and provide an excellent basis for biometry. Moreover, measurement of gross dimensions of the shell at successive ontogenetic stages is a practice widespread among fusuline taxonomists. There is a vast accumulation of data on diameter of the proloculus and on half length of the coiling axis, spiral radius and radial diameter of chambers (Fig. 21) at each volution. Dunbar and Skinner (1937) and Thompson (1954) are examples. Indeed, fusuline workers appear to have made the most consistent and persistent efforts to measure the form of the foraminiferal shell at successive stages of its growth. This expresses the realization (Dunbar and Skinner, 1937) that many species change proportions during growth and that such changes are useful in taxonomy.

Data collection problems concern orientation of sections (Cutbill and Forbes, 1967) and identification of volution number on which measurements are made (Burma, 1942). Variance of measurements tends to be greater in axial than in sagittal sections (Burma, 1948). But the overriding problem of fusuline biometry is that of representation and analysis of the ontogenetic data that has been collected. By one method, a measure of shell size attained at a given number of volutions of the spiral is plotted against the latter number. Comparison of shell dimensions at the ith volution are then made (Burma, 1942). Erroneous conclusions can be drawn from this method if the size of the proloculus is variable in the populations sampled (Dunbar and Skinner, 1937). Ozawa (1970) showed the magnitude of variation in megalospheric proloculi may be very large. Generally, the dimensions of the first chamber are a function of the size of the proloculus. At k volutions the dimensions of a microspheric shell are likely to be smaller than at k volutions of a megalospheric shell. Volution number provides an unsatisfactory reference point for comparison of measurements, even if rate of chamber expansion is constant in both micro- and megalospheric individuals, because it does not ensure that the origins of the spirals are at a common locus. Douglass (1970) demonstrated the influence of volution number in a comparison of measurements of two samples of *Monodiexodina kattaensis* (Fig. 22).

Two solutions to this problem of representation, and comparison, of measurements of spiral form have been used. Roberts (1953) plotted measurements (say, half length: spiral radius) as bivariate data. This is a relative growth solution. Neither variate is dependent. The effect of volution number is removed because the components of each (x, y) pair refer to the same volution. A growth curve technique was demonstrated by Cutbill and Forbes (1967; see also Douglass and Cotner, 1972). Cutbill and Forbes compared plots ("spiral curves") of a shell dimension (dependent variate) against volution number (independent variate) for several specimens by superimposing them at selected loci.

The relative growth approach is satisfactory provided that regression statistics are computed for individuals as well as for the whole sample. Roberts failed to compute the former, thus intrasample variability was not assessed. Techniques suitable for analysis of variance in ontogenetic data are described by Cock (1963). Cutbill's and Forbes' procedures permit only graphical comparison of data. They are arbitrary in that the shape of variation envelopes tends to vary according to the locus selected for superimposition. They provide no assessment of the signifi-

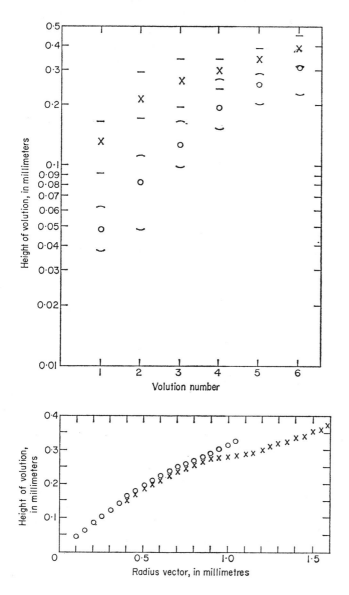

Fig. 22. (Upper) Marked separation of plots for two samples of *Monodiexodina* when height of volution is plotted against volution number. Mean values and observed ranges are shown. (Lower) Same ordinate, plotted against radius. (After Douglass, 1970).

cance of variation in slope of curves among individuals. Neither does the computational procedure of Douglass and Cotner (1972). Here again, analysis of variance techniques for ontogenetic data should be suitable, perhaps with transformation of data, but have not been applied. If fusuline data have been under-analysed, it is not because of the lack of suitable methods.

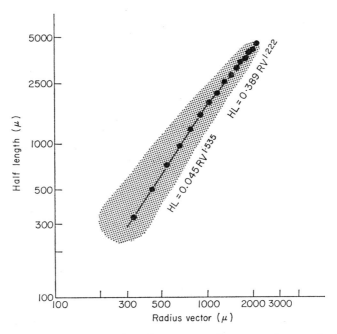

Fig. 23. Mean half length plotted against mean radius vector in growth of shells of *Colania douvillei*. Variates show allometric relationships with change in slope at seventh chamber corresponding to structural modification. (After Ozawa, 1970).

For some fusuline taxa, shape tends to be constant during ontogeny as judged by absence of trends in form ratios (half length:radius), volution to volution. If the spiral is logarithmic with log $r = Cv$ (r = spiral radius, v = volution number and C is a constant that is fixed throughout ontogeny), then spiral curves on semilogarithmic coordinates will be linear. Cutbill and Forbes termed such linear curves "normal". For the *Schwagerina* and *Paraschwagerina-Pseudoschwagerina* complexes they found that spiral curves for shell diameter decline in slope in later stages of growth. Ontogenetic change in shape occurs. Ozawa (1970) investigated this in *Colania* (Fig. 23). Two growth constants were required to fit ontogenetic data (averages) for half length and spiral radius. Change in slope coincides with appearance of the secondary transverse septula. Koepnick and Kaesler (1971), from analyses of shell characters at standardized volution numbers, found that all but septal count were directly size dependent. They considered that an inverse relation between septal count and shell size may be characteristic of fusulines. But this conclusion was made from samples that were cross sections of many ontogenies. Analyses of septal counts within each ontogeny and of frequency distributions of prolocular diameters are needed to explicate the relation.

2. NUMMULITIDAE

The coiling axis, in contrast to fusuline taxa, is short and the shell is discoidal rather than fusiform. Because of this form, equatorial sections have been the major source of biometrical data. Spiral diameter and

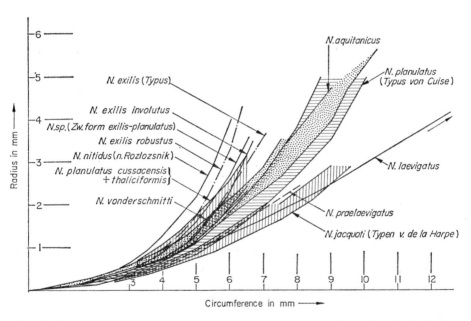

Fig. 24. Expansion of radius, volution by volution, in growth of *Nummulites*. Individual ontogenies as lines, range for several ontogenies hachured. (After Schaub, 1951).

length of the shell, number of septa and radii of successive revolutions of the spire were collected for many taxa by Rozlozsnik (1929) and Schaub (1951). Ontogenetic data on growth of radii were presented graphically in spiral diagrams (Fig. 24). Arni (1967) expanded the representation to include other diagnostic ontogenetic data. On arithmetic coordinates the spiral diagrams (Rozlozsnik, 1929; Schaub, 1951; Haynes, 1962; Blondeau, 1965; Arni, 1967) show increment in slope in early ontogeny but some subsequently inflect to produce an overall sigmoidal shape. Decrease in slope in late ontogeny is probably related to change in chamber shape (Schaub, 1963) as is change in the spiral angle during growth (Carter, 1953; Mangin, 1966).

Khan and Drooger (1971) objected to the conventional spiral diagram because of difficulty in assessing significance of differences among curves. Indeed, this seems not to have been attempted statistically. Khan and Drooger (1971) and Drooger *et al.* (1971) therefore abandoned the conventional data set and spiral diagram of Roslozsnik and Schaub and studied prolocular diameter, spiral diameter of the first two revolutions and the number of chambers formed in this distance (Fig. 25). The latter

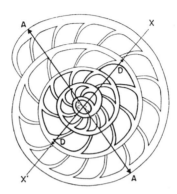

Fig. 25. Measurements applied to equatorial sections of nummulite species by Khan and Drooger (1971) and Drooger *et al.* (1971). The number of chambers in first two revolutions is counted. (After Drooger *et al.*, 1971).

two variates, at fixed ontogenetic position, provide a type of static data (chapter IV). They could also be obtained as a subset of a full ontogenetic set of data on spiral diameter and chamber number. The data collected by Drooger and his colleagues is simple to analyse and permit tests of significance. However, they certainly reveal much less about ontogenetic development than the spiral diagrams of previous workers. These, like comparable fusuline data, could well be analysed using the model of Cock (1963). I suggest that there is an urgent need to experiment with this model using ontogenetic data from groups such as fusulines and nummulites. It is a retrograde step to cease collection of full ontogenetic data on the premise, probably false, that they are difficult to analyse.

3. ALVEOLINIDAE

Parallel trends towards axial elongation of the shell in lineages of *Alveolina* were mapped by Hottinger (1960) using the ratio between axial and spiral diameters. There is rapid increment of the ratio in early stages of growth. Thereafter, the ratio stabilizes or may decline. There are specific contrasts in the shape and location of ontogenetic plots of the ratio (Fig. 26).

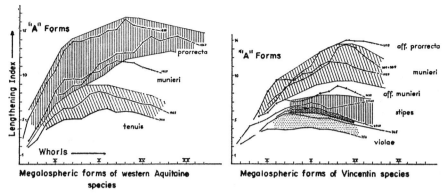

Fig. 26. Ontogenetic history of shape of shell in *Alveolina*. Ordinate is ratio between axial and spiral diameters of shell. (After Hottinger, 1960).

4. OTHER PLANISPIRAL AND ALLIED TAXA

Most measurements applied to external features of planispiral shells with short coiling axes are equivalent to those used on their multiserial turbinate counterparts. Thus characters such as greatest spiral diameter, spiral diameter at 90° to the greatest, length of the shell, and number of chambers in the last revolution have been used by Ellerman (1960) and Buzas (1966) on *Elphidium*, by Maync (1959; 1960) on *Pseudocyclammina*, Theyer (1971) on *Cyclammina* and Berggren *et al.* (1967) on *Pseudohastigerina*. Only the measurement of diameters of dextral and sinistral umbilici in the latter work introduces a character without an equivalent in turbinate shells.

Two spiral diameters and length can be measured on shells in which the planes of coiling of successive revolutions rotate. Wolanska (1959) considered the shell of *Agathammina pusilla* as an ellipsoid and used these three measurements as estimates of axes. Shells were classified according to their ellipsoidal shape.

C. Has Measurement Supplanted Words and Pictures?

My survey of biometrical studies of the foraminiferal shell is partial but sufficiently representative to allow evaluation of this question. Patently, the answer is "no". The fundamental reasons are, firstly, that quantitative characterization of chamber shape has scarcely commenced. The chamber is the building module of the shell and its shape, at successive stages of ontogeny, is a crucial component of shell form. Secondly, if chamber shape is neglected, then the usual application of two or three dimensions to capture the form of the shell is grossly inadequate. An informative occupation is to draw a perspective of three orthogonal axes corresponding to length and spiral diameters and sketch the variety of shell form that they can accommodate. The experimenter will soon find need for information on spiral geometry and on chamber shape if he wishes to describe even the outline of a particular shell.

Why have data sufficient to characterize shell form not been collected? The low-power stereomicroscope certainly provides sufficient resolution for many characters. One explanation is that biometricians have aimed at supplementing "words and pictures", not supplanting them. Thus we find biometricians working within the confines of established qualitative taxonomies, perhaps testing them with quantitative data, but not attempting to classify anew. Studies of fossil taxa show a parallel tendency. Characters that exhibit directional, evolutionary, changes useful in stratigraphy have been searched for, to the neglect of others. Research into nepionic chambers in larger foraminifera provides an example. Note how a simple enumerative approach yielding useful stratigraphical results has been adopted while chamber shape and shell geometry remains largely unexplored. The failure to analyse ontogenetic data, even where they are extensive as with fusuline taxa, further exemplifies the thesis that biometry is an adjunct, perhaps only a minor adjunct, to long-established qualitative methodology in taxonomy.

Another explanation is that foraminiferal biometry has been heavily constrained by problems of data capture and analysis and that until these problems are resolved, qualitative methodology must continue to predominate. Some facets of this argument are now explored.

IV. Data and Analysis

This survey is not restricted to techniques of analysis that have been applied to data on the foraminiferal shell. Biometricians concerned with foraminifera have usually applied simple, well known, statistical methods. This is justified as most workers are zoologists, paleozoologists or biostratigraphers, not statisticians. However, this situation may lead to under-analysis of data and justifies several of my excursions (as a biostratigrapher) into methodology largely ignored in published research on foraminifera.

A. Collection

Eyepiece scales and filar micrometers remain the tools of trade of the foraminiferal biometrician. They provide reproducible data and are limited primarily by the associated optical system. For measurement of large fusuline shells Sanderson (1971) used stage micrometers. These obviate the necessity for several partial measurements when the dimension to be recorded is larger than the field of view. For small structures, say with dimensions of less than 0.01 mm, light stereomicroscopy provides insufficient resolution for measurement. Here scanning electron microscopy promises a major advance for biometricians. Olsson (1971,

1972) reported biometrical studies based on measurements collected from micrographs. But as yet the potential, intimate, connection between advances in scanning electron (and X-ray projection) microscopy and biometry is unrealized by foraminiferal workers. These microscopes produce large increases in the amount of information obtainable from the foraminiferal shell. Hay (1971) found that the amount of data potentially available from a scanning electron microscope was 400 times that from a stereoscopic binocular light instrument. Yet neither he nor Sylvester-Bradley (1971) envisaged that biometry would be required to process the additional information. Indeed, Hay considered methods for reducing the amount of image information initially collected by the instrument. If the data collected are excessive (for visual applications) but not redundant, then surely the requirement is for quantitative methodology to analyse and compare the analogue signals produced by the microscope.

If microscopy is adequate for most structures of the foraminiferal shell, then collection of measurements is not. Manual recording of dimensions is slow and fatiguing. Recording errors, in my experience, occur frequently. Moreover, if processing is by computer, then the manually recorded data have to be either manually punched on to computer readable media or read automatically by an optical character device (Alexander-Marrack, Friend and Yeats, 1971). The latter usually requires standard character sets and the biometrician's original worksheet is unlikely to be satisfactory. Data loggers provide a practical solution to the problem of collecting measurements from the microscope. In their simplest form these instruments convert the coordinates of a locus on the object studied into digital form suitable for computer processing. On a commercial instrument used by Piper (1971) for grain shape studies, the loci are selected by the operator using a reading pencil. This instrument is suitable for photographic enlargements or projections of foraminiferal images. I use a stereomicroscope with transducers linked to the mechanical stage. The loci delimiting a linear measurement, say chamber diameter, are selected by stage manoeuvring, their stage coordinates recorded automatically on paper tape, and a simple computer program reconstructs the measurement from the coordinates. Walker and Kowalski (1972) described a data bank for x, y coordinates (human craniofacial data) logged by a simple, manually operated, instrument. Interactive access to a computer with facilities to visually display data is highly desirable.

Simple data loggers of the type described eliminate manual recording of measurements and markedly reduce time required to collect data. Some of the time saved may profitably be employed to replicate measurements so permitting operator error in locating measurement loci to be assessed. With manual recording this is performed only exceptionally.

But, in longer perspective, these devices appear to be only a first rather primitive step in the collection of data from the foraminiferal shell. In other fields of biological research good progress has been made with automatic scanning of images. Neurath et al. (1966) adapted a TV camera to serve as a scanning device for collection of data on chromosome morphology. Ledley et al. (1972) surveyed this field. Prewitt and Mendelsohn (1966) described a project for classification of leukocytes in which cell images were recorded by a flying spot scanner while Feldmann and Bryan (1970) used a computer controlled microscope and digitized transmitted light to 256 levels. The optical density of the object was measured by some of these instruments at intervals less than a micron apart. Hawkins (1970) discussed principles. Sophisticated systems have been developed to reconstitute the image, delineate structures, analyse their form (Klein and Serra, 1972) and perform classificatory functions. At least one system is currently available commercially. Picture processing languages have been investigated (Narasimhan, 1970; Feldman and Bryan, 1970). To most foraminiferal taxonomists these developments may seem to be quite irrelevant. To some they may seem to usurp the taxonomist's function. To the biometrician they promise to revolutionize data collection and processing. It is a fair prediction that if biometry is to become a technique of wide relevance in foraminiferal taxonomy then automatic collection of data will be essential.

B. Operator Error

Errors in measurement of a structure arise from procedures followed during preparation of the material (orientation of section or of free specimen) as well as in selection of measurement loci. The subject is seldom explicitly discussed by foraminiferal biometricians and rarely analysed. Wrongly located sections are an obvious source of error (Dunbar and Skinner, 1937). Drooger and Freudenthal (1964) noted that the value of van der Vlerk's factor A (Fig. 3), measured on embryonic chambers of lepidocyclinids changed from 43 to 50 in the course of fully exposing the nepionic chambers. The latter were taken to define a reference plane.

Reproducibility checks may be performed by the same operator. Drooger (1955) recounted the number of nepionic chambers in *Cycloclypeus* after a lapse of 3 months and only those specimens for which counts were identical were included in the sample data. More stringent tests involve several operators. Smith (1963) simply noted that measurements of gross shell dimensions of *Bolivina* by another operator seldom varied by more than one micrometer division. But tests of significance of differences between operators are required if the measurements are proposed for general use among researchers. An analysis of variance model for two-way classification with replications is adequate for this purpose. Griffiths (1967) discussed such a model in sedimentological contexts and provided sufficient information for paleontological applications. Scott (1965) applied this model to assess operator variation in measurement of chambers in a textularid lineage. The principal difficulty confronting operators was location of the suture between chambers. In agglutinating taxa this is often obscure. Operators were consistent from specimen to specimen (Fig. 27) and from day to day in repeating

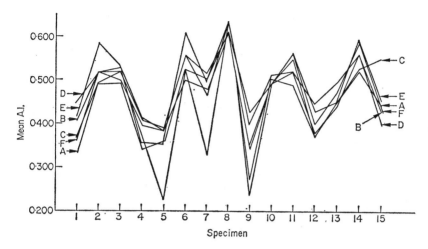

Fig. 27. Variability among operators A–F in measuring an index (A.I.) of uniseriality in *Haeuslerella*, a textularid genus. Degree of consistency indicated by parallelism of lines. (After Scott, 1965).

measurements. However, there were significant differences between mean values recorded by the six operators. As the measurements were intended to map the progress of a trend in chamber arrangement for stratigraphical purposes, this result implied that values obtained by different operators could not be directly compared. Identical values for the trend from two operators would not necessarily indicate populations at the same stage in the trend. Note that the experiment was not designed to assess variability among operators in positioning specimens for measurement. To my knowledge no set of measurements on the foraminiferal shell proposed for stratigraphical or taxonomic applications has been rigorously tested for all sources of operator error. This is certainly a valid reason for others to neglect existing biometrical studies.

C. Univariate Methods

Foraminiferal shell structures have been analysed chiefly by simple univariate methods for taxonomic ends. Univariate statistics, and the limitations imposed by the models from which they derive, have been widely discussed. Of equal importance to the biometrician are limitations imposed by the type of data, the nature of the sample and by the biological implications of the structures measured. In taxonomic studies, particularly, effects due to growth of structures must be identified. For certain univariate data this requires controlled sampling.

1. LINEAR MEASUREMENTS

In general, measurements of a structure provide size information. Foraminiferal biometricians deal with a shell which is an aggregate of chambers added successively during growth. What information is represented by statistics for a single variate? The frequency distribution of a gross shell dimension for a sample from a living population reflects the age structure of the population. For a death assemblage it reflects, in addition, the differential effects of mortality (predators, disease) on the age structure. Still more factors (transport, selective preservation and extraction) may influence the distribution for a fossilized sample. Finally, there may be geographic or stratigraphical trends in the age/size structure of populations. Thus, although statistics have been compiled for a gross dimension of the shell (Smith, 1963; Kennett, 1968b; Hansen, 1970), their taxonomic interpretation may be complex (Lewis and Jenkins, 1969). One step towards elimination of unwanted variables is to sample gross shell size at some standard stage in ontogeny. But this procedure, too, has problems. It is exceptional to have information (in time units) on age of specimens. How then to identify stage in ontogeny? Recourse to a specified number of chambers or whorls (Drooger et al., 1971) may lead to spurious results if polymorphism occurs but is not detected. The size of the foraminiferal shell, at a given number of chambers tends to be correlated with the size of the proloculus. Indeed, the initial chamber marks the growth stage least ambiguous to identify and most suited to univariate analysis. Provided that microspheric and megalospheric groupings can be correctly identified size measurements of this chamber are likely to provide data least influenced by variations in age structure and mechanical processes in the environment (both life and post-mortal) that operate differentially on gross size of the shell.

2. ENUMERATION DATA

The direction in which a turbinate foraminiferal shell coils exemplifies binary enumeration data in which ontogenetic influence is absent. To the contrary, number of chambers is a function of stage in ontogeny. In the first example, ontogenetic restrictions on the sampling procedure are unnecessary. But note that, if one direction had strong selective advantage during growth, the proportion of shells coiled in that direction could vary according to the age structure of the sample.

The number of chambers formed in early ontogeny has been studied extensively in larger foraminifera. As the samples studied seem always to consist of specimens in which post-embryonic chambers are present, the data are not exposed to variability due to stage in ontogeny. Less fortunate are studies that involve number of chambers formed in late ontogeny. Grabert (1959) enumerated the number of biserial chambers in Cretaceous lineages of *Gaudryina* and *Spiroplectinata* and showed that dramatic extension in number of these chambers occurred. That the data portray phyletic trends cannot be doubted. But dependency of the upper frequency polygons (Fig. 15) on the type of sampling scheme is shown by their truncation at four biserial chambers.

3. RATIOS

Use of a ratio permits several variates to be analysed as one. This provided computational convenience when techniques and machines for processing multivariate data were not available. Again, to the taxonomist, a ratio of linear measurements may be envisaged as a shape variate that is at once more readily interpretable and informative than any of the components of the ratio treated singly. Thus the ratio of half length to spiral radius ("form ratio") is conventionally tabulated by fusuline taxonomists while analogous ratios have been applied to many smaller foraminiferal shells.

Are ratios suitable for analysis of bivariate, sometimes multivariate, data? Consider the bivariate case of two linear measurements of the shell. As a ratio they provide a dimensionless number that characterizes an aspect of shell form. But an immediate disadvantage is that the ratio obscures information about sources of intrasample variability and co-variation of the characters. Especially important in taxonomy is the question of size or age correlated changes in shape. Does the ratio change with increase in size of the shell? As a single variate the ratio gives no clue. Such information is forthcoming only if the ratio is plotted against a measure of shell size or, more directly, is dismembered and plotted as simple bivariate data. Commonly, the variate representing shell size is one that is already a component of the ratio (Smith, 1963; Klaus, 1960). This is a circuitous method for investigating the influence of size on shape. Sokal and Rohlf (1969) noted the relative inaccuracy of ratios and problems with their distribution but, to the biometrician studying intrasample variation, the suppression of information on covariation of the variates forming the ratio is the outstanding disadvantage. In this respect their use is indefensible.

4. RELATIVE VARIABILITY OF SHELL STRUCTURES

Univariate measures of location and variance are used to assess significance of differences of single characters between samples. But biological interest also centres on comparisons of intrasample variability.

A dimensioned measure such as variance or standard deviation is unsuitable for this purpose because there is a general tendency in biological populations for variance to be a function of the mean. The coefficient of variation (Simpson *et al.*, 1960) which is the quotient of sample standard deviation and mean expressed as a percentage, is widely used by biologists as a measure of relative variability. Lewontin (1966) showed that the variance of logarithms of measurements provides a measure of relative variability that is more convenient, computationally, when tests of significance are made. The standard deviation of $\log_e (x)$ is proportional to the coefficient of variation of variate x.

The coefficient of variation permits examination of several questions. Are certain characters of the shell more variable than others? Are there intertaxon differences within the order in variability of the same or analogous structure? Is the foraminiferal shell highly variable in comparison with other organic structures? Many others could be posed but even of those listed only partial answers are possible. Values of the coefficient of variation (V) for spiral diameters of chambers formed late in ontogeny of *Globigerina* and *Globigerinoides* (Berggren and Kurtén, 1961; Malmgren, 1972; Scott, 1970) lie in the range 10–23. For apertural height in *Globigerinoides* (Scott, 1970) values of V as high as 72 were obtained. Spine length in *Asterotalia* gave $V = 40$ (Ghose, 1966) while for the greatest spiral diameter of the shell the value was 16. A particularly high value of 181 was obtained (MacGillavry, 1965) for the number of adauxiliary chambers in *Lepidorbitoides*. Cosijn (1942) recorded $9 \leqslant V \leqslant 45$ for shell, protoconch and deuteroconch dimensions in various larger foraminifera and noted that, in general, values were greatest for the shell and least for the protoconch. In some lineages there was a tendency for values to decline in the course of evolution. Smith (1963) obtained $V > 50$ for shell length in some bolivinids but the wide ontogenetic range sampled probably contributed to such high values. Values obtained by Ozawa (1970) for spiral radius, half length and the ratio of these two measurements in a fusuline species are particularly valuable because they relate to successive volutions in the growth of the shell, with megalospheric individuals segregated from microspheric. There is a pronounced trend for V to decline from about 20 at the first volution to less than 10 at the conclusion of growth (Fig. 28). The standard deviations of the measurements on which V is based tend to increase less rapidly than their mean during growth of the shell. There may be several explanations for this but one that merits further inquiry is that selection operates on shell characters more intensely in later stages of growth than in earlier stages.

MacGillavry (1965; 1971) developed Cosijn's investigation into values of V for the same structure in various taxa. His choice of the proloculus (megalospheric where differentiated) is wise in view of the problems, already discussed, that hedge the interpretation of univariate statistics on linear measurements of the foraminiferal shell. MacGillavry fitted a linear function to data on mean prolocular diameters and their standard deviations (Fig. 29) from taxa that in some instances were only very distantly related. The slope of the fitted line was taken as a "group index of variability" and corresponded to a coefficient of variation of 19.

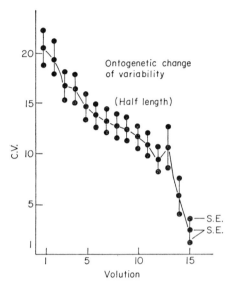

Fig. 28. Reduction in coefficient of variation (c.v.) during growth of the shell of *Colania douvillei*. S.E. is standard error of c.v. (After Ozawa, 1970).

This research is a valuable contribution towards the definition of a standard of comparison for relative variability studies of the foraminiferal shell. But it must be recalled that influences on the frequency distribution of measurements of shell structures are many, especially when fossilized death assemblages are sampled, as with some of MacGillavry's data. Investigation of V for the proloculus in living populations of foraminifera of known age structure appears to be the next step required in this research.

Coefficients of variation for foraminiferal structures commonly lie between 10 and 30. Values greater than 30 sometimes occur and the sources of such high relative variability require detailed investigation. Some structures (spines, apertures) may exhibit greater intrinsic variability than say, chamber dimensions, although data that stringently test this suggestion have yet to be collected. Equally interesting is the apparent lack of values less than 5. In skeletal structures of organisms as diverse as vertebrates (Simpson *et al.* 1960; Simpson, 1953) and

Ostracoda (Reyment, 1960) such values are common. Some of these samples refer to accurately defined age segments of populations. In many foraminiferal samples age structure is quite unknown. However, this factor does not account for the magnitude of V for the foraminiferal proloculus. Similarly, the fact that most values for foraminifera are from fossil samples is not, in itself, an adequate explanation as Reyment's data for Ostracoda also relate to fossils.

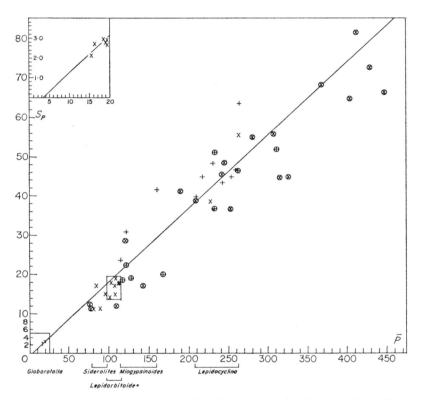

Fig. 29. Plot of standard deviation (ordinate) against mean diameter of megalospheric proloculus for *Globorotalia* and various larger foraminifera with line of best fit. $+$: turbidites; $\times$: non-turbidites; $\otimes$, $\oplus$: Nummulitidae. (After MacGillavry, 1971).

Although comparisons with metazoan structures suggest that the foraminiferal shell is highly variable, data for some structures produced by other simple or unicellular organisms are required to obtain perspective. For the number of plates in the external series of the superior disc in the Jurassic coccolith *Ellipsagelosphaera frequens* (raw data from Noël, 1965, p. 126; $n = 15$) the coefficient of variation is 16. Variability of the dimensions of the superior disc along the major and minor axes of the coccolith is much higher ($V = 44$; $V = 38$). For measurements of the marginal crown in *Zygolithus erectus* (Noël, 1965, p. 64; $n = 11$) the range is $21 \leqslant V \leqslant 28$. Values for diameter of distal shields and for number of segments in the corresponding shield in *Cyclococcolithina leptopora* (raw data from Ellis *et al.*, 1972, text-figure 5; $n = 49$) also lie in this range. Jerković (1969) tabulated measurements for gross skeletal dimensions of the Neogene silicoflagellates *Dictyocha šoljanii* (p. 46; $n = 51$), *D. slavničii naviculoidea* (p. 51; $n = 48$) and *Deflandryocha intercalaris* (p. 61; $n = 49$); coefficients lie in the range $14 \leqslant V \leqslant 25$ with the exception of a lower value ($V = 7$) for length in *Dictyocha slavićii naviculoidea*. Jerković recognized several subspecies of *D. slavnićii* from one locality. Intertaxon discrimination may have influenced the latter result. This survey is certainly insufficient to support any generalization but it does suggest that intrinsic variability of skeletal structures formed by other simple organisms is comparable with that exhibited by the foraminifera.

D. Shape and Size

Shape is of crucial importance in foraminiferal taxonomy at infrafamilial levels. Quantitative characterization and analysis of shape is perhaps the most important contribution biometry can offer to the study of the foraminiferal shell. In intrapopulation variation, the relation between size and shape is relevant to taxonomy and, more fundamentally, to the explanation of shape. The ratio of two characters of the foraminiferal shell provides a shape variate that is unsatisfactory because size information is suppressed. In contrast, if the characters forming the ratio are treated as separate variates then even a simple bivariate scatter provides explicit information on the relation between size and shape. Bivariate methods thus form the foundation of analyses of shape. Although studies must involve at least two variates, there is no intrinsic reason to restrict investigations to pairs of variates. The present emphasis on bivariate analysis reflects in part the state of theoretical advance by mathematical statisticians and also the level of practical knowledge by biometricians. However, development of multivariate techniques is unlikely to greatly diminish the importance of bivariate studies, if only because of their graphical simplicity.

1. DATA CATEGORIES

Types of data used in shape studies should be carefully distinguished. There is a fundamental dichotomy between data collected from the same, or equivalent structure, on one individual at successive stages of its growth and those comprising one measurement set per individual with, or without regard, to stage of growth. The categories discussed here are adapted from Cock (1966) and elaborate this dichotomy.

(*a*) *Static Data*

A set of measurements is applied once, and only once, to each individual in the sample. Individuals are uniform either in age (in time units) or stage in ontogeny. Because of this uniformity, the data are not relevant to the study of ontogenetic changes in shape. Cultures of known age (Sliter, 1970) could provide data of this type. In dead and fossil material one can standardize for chamber number, on the assumption that equality in chamber number indicates similarity in ontogenetic stage. This may be erroneous if the taxon is polymorphic.

(*b*) *Ontogenetic (or Longitudinal) Data*

A set of measurements is applied at successive growth stages or known times in the life of the individual. Such data can provide a full record of the growth of individuals. Subsets may be extracted to provide static data as defined above, or cross-sectional data (Cock, 1966) which consist of a series of subsets of static data for individuals at various ages. The foraminiferal shell, formed by accretion of successive newly formed chambers to the already existing structure, provides a full ontogenetic record of its form. With involute plane spiral shells and turbinate shells, sections are necessary to reveal the record in certain planes. In groups in which taxonomy has been developed on the form of the shell in section, much ontogenetic information is available. Quantitative data has been collected mainly in the Fusulinidae, where chambers regularly expand in size throughout ontogeny. In contrast in taxa, such as *Cycloclypeus*, in which the regular spiral sequence is greatly reduced, quantitative data relates only to the earliest stages of shell construction up to the termination of nepionic chambers. As previously emphasized, very little analysis of the data as longitudinal data has been made.

(*c*) *Mixed Data*

One set of measurements is obtained from each individual. Standardization of the sample for age, or stage in ontogeny is either neglected or cannot be rigorously applied. Almost all bivariate studies of smaller foraminifera have utilized mixed data collected from shells at conclusion of construction. Lack of standardization for age means that the data may be drawn from several (unrecognized) static subsets. Their number depends on the segment of the age distribution of the population that is sampled. Mesh sizes in sieves used to prepare foraminiferal residues establish arbitrary lower size limits.

Mixed data of the type here described do not contain information about the history of shape in any one individual. Rather, the information relates to shape of individuals of various sizes at conclusion of growth of the shell. An essential point is that, depending on the sampling scheme and the age structure of the populations, shells at various stages of growth will be represented and some information on growth is thus likely to be present. The problem is to estimate ontogenetic information from data that represents only one locus in the ontogeny of each shell.

2. ANALYSIS OF BIVARIATE DATA

Relationship between pairs of shell measurements is commonly analysed by techniques of linear regression. The method of fitting a straight line depends on the nature of the data and the intent of the investigator. There are two major cases.

(a) Functional Relationship

Consider an investigation, using cultured populations, into the relation between variation in length of the shell (l) and age (t). At selected times, length is measured. Time is the independent variate (conventionally plotted as the abscissa) and is deliberately selected. To the contrary, there is no control exercised over length (the dependent variate) and it may be considered to vary randomly. The least squares method (Sokal and Rohlf, 1969) is appropriate for fitting a line $l = a + b_{lt}t$ to such a scatter, if linear. The line provides an efficient estimate of the functional relation of l to t and may be used to predict l, given t. The regression coefficient b_{lt} gives the increment in l for unit increment in t. Note that with such data the regression of t on l cannot be correctly computed: that is, to estimate age from length.

(b) Structural Relationship

Here the objective is to obtain a line which best fits the data. This is very commonly required in foraminiferal biometry. Investigators have tended not to seek to explain variation in one character by studying its regression on another. Rather they have used a line of best fit to represent the orientation of a bivariate scatter, particularly in intersample comparisons. Commonly, both variates may vary randomly. The least squares approach is unsatisfactory for representing structural relationship. It provides two lines $y = a_y + b_{yx}x$ and $x = a_x + b_{xy}y$ and these are coincident only in the exceptional case in which all points in the scatter are colinear.

Two methods for computing a line of best fit are in common use. Imbrie (1956) described paleontological applications of the reduced major axis method. Ghose (1970) discussed its performance with data on *Pellatispira*. He found that the line was unbiassed between the variates and suggested that it could be accepted as a compromise between the two least squares regressions. Simpson *et al.* (1960) pointed out that a principal disadvantage of the technique is that exact confidence intervals cannot be calculated for the coefficient b. In this regard, a second method

(Bartlett, 1949) for representing a structural relationship between two variates, is advantageous. It is also more satisfactory than the reduced major axis method when correlation between the variates is weak. Computations are set out in Simpson *et al.* (1960) and Ghose (1970) described its utility with foraminiferal data (Fig. 30).

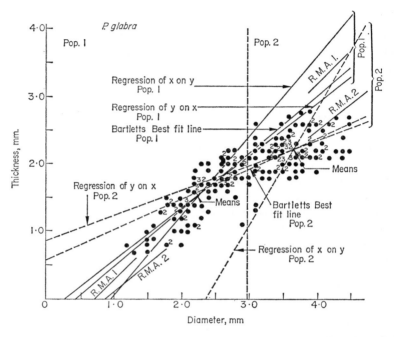

Fig. 30. Reduced major axes, least squares regression lines and Bartlett's best fit lines for two samples of *Pellatispira glabra*. (After Ghose, 1970).

It is apparent from the discussion of data types that most information is provided by longitudinal sets. Estimates of slope and position of can be obtained for:

    i. each individual ontogeny
    ii. a single stage in ontogeny
    iii. a set of stages in ontogeny
    iv. pooled data

Cock (1963) described a model for analysis of slopes of lines fitted to longitudinal data that could well be applied to foraminifera. Studies so far reported of longitudinal data from foraminifera have either not progressed beyond graphical representation or have failed to consider variation in individuals during ontogeny. Thus Ghose (1970) fitted lines to means at several stages of ontogeny, a form of cross-sectional analysis while Roberts (1953) pooled data from several individuals. Information on variation in slope among individuals is suppressed. Similarly, it is suppressed in the graphical representation by Douglass (1970). It is likely that, even under transformation, certain foraminiferal data will be non-linear. However, those segments of ontogeny that are representable by linear functions may still be analysed by Cock's model.

3. ALLOMETRY

Gould (1966, p. 629) viewed allometry as the study of size and its consequences or, more explicitly, as "the study of proportion changes correlated with variation in size of either the total organism or the part under consideration". Allometric studies of the foraminiferal shell are few (Olsson, 1971) but the significance of the subject should not be underestimated. Bertalanffy (1960) observed that allometry relates to the principle that the organism must remain functioning and self-maintaining in spite of variations in size. Thus it directs attention to concepts such as "function" and "adaptation" that are central to the interpretation of shape and of changes in shape, whether they be onto-genetic or phylogenetic. In turn, I believe this impinges beneficially on the quality of classifications. To the contrary, McGowran (1971) considered that our ignorance of the functions of the foraminiferal shell and of functional adaptation is not strictly relevant to taxonomy. This view I reject. Should we not understand, as fully as possible, the functions of the structures upon which classifications are based? Pratt (1972a) evaluated classification methodology and remarked that it is from engineering knowledge of the organism that the biologist selects characters likely to be general in the taxon.

The allometric formulation assumes that the growth rate (dy/dt) of a character is multiplicative, that is, proportional to the existing magnitude of the character. Thus $dy/dt = K_1 y$ where K_1 is the growth constant for the character. Similarly for another character (x), $dx/dt = K_2 x$. Then the relation between the variates is $y = \beta x^\alpha$ where α, the constant of allometry, equals K_1/K_2 and β is the y-intercept at $x = 1$. White and Gould (1965) showed that β is a factor depending on scale and discuss its biological interpretation. Now if α is constant throughout ontogeny (longitudinal data) or throughout the size range sampled (mixed data), then a linear scatter results when data are plotted on logarithmic coordinates. Identity of proportions throughout the data range is indicated when $\alpha = 1$ signifying that the two growth constants are equal. This special case is termed isometry. "Negative allometry" and "positive allometry" refer to variates for which ratios of the growth constants are respectively less than and greater than 1. For a rectilinear scatter on arithmetic coordinates, isometry is indicated when a fitted line intersects the origin. Only when $a = 0$ in the expression $y = a + bx$ is y/x constant for all values of x. Gould (1966) gave an instructive example showing how increments of $y/x = 2/1$ for each unit increase in x transforms a square into a rectangle. The increments are constant and the line is linear but there is continuous change in shape of the figure.

Ghose (1970) suggested that the allometric formula is inapplicable to most foraminiferal shells because their growth is accretive, not multiplicative. If multiplicative growth does not occur then biological bases for the formula are removed. This opinion requires examination. It is correct, of course, that the foraminiferal shell accretes, chamber by chamber, during growth. However, the question really concerns the form of the relationship between the sizes of successive chambers. Research on rate of expansion in some multiserial taxa, reviewed earlier, suggests that diameters of successive chambers d_i, d_{i+1}, are approximated by $d_{i+1}/d_i = k$. Normally $k > 1$ although there are exceptions in the later growth of some planktonic taxa and probably also in certain fusuline and nummulite taxa. I have not seen evidence for an additive relation such as $d_{i+1} = d_i + k$ ($i = 1, 2, \ldots, n^{\text{th}}$ chamber). Here d_{i+1}/d_i will decline throughout the sequence. At this stage there seems to be fair empirical evidence to support a multiplicative relation for the chamber suite. Ghose equates accretive growth with a linear relation between variates. But a linear relation can arise with multiplicative growth when the growth constants K_1, K_2 for each character are the same. This occurs in foraminifera with two measurements on chambers when chamber shape is the same for successive growth stages. Again, provided chamber shape is constant, certain measurements of the shell, such as greatest spiral diameter and the spiral diameter at 90° to the greatest, will exhibit a linear plot (ontogenetic data) whether or not chambers expand at a constant rate over the last volution of the spiral. The two measurements of the shell primarily reflect the growth behaviour of a single chamber variate. Beyond this, there is the general condition that linear relations between variates (arithmetic coordinates) signify allometry whenever the regression line does not intersect the origin. Thus linear relation between variates may indicate either isometry or allometry without implying that growth is other than multiplicative.

Foraminiferal studies are remarkable for their neglect of size correlated changes in shape. In orbitoidal and similar forms, study has concentrated on counts of chambers (e.g. nepionic acceleration) and changes in shape during ontogeny either of the chambers or of the shell remain unquantified. Similarly, with the many convergent trends towards reduction in number of chambers per revolution in smaller foraminifera, biometricians have worked on simple indices to map the trends rather than investigating change in shape, whether ontogenetic or phylogenetic. Gould (1968) observed that questions about changes in shape provide a valuable methodological approach to adaptive explanations of form. Foraminiferal biometricians have not asked these questions and, as a consequence, biometry has contributed little to the explanation of shell morphology.

4. CORRELATIONS AMONG CHARACTERS

The product moment correlation coefficient measures the intensity of the relation between two variates. If we have n values of variate i and of variate j, these may be envisaged as two observation vectors in n-dimensional space. The cosine of the angle between the vectors is equivalent to the correlation coefficient r_{ij}. Olson and Miller (1951, 1958) represented organic structures by a set of linear measurements and showed that the matrix of correlation coefficients provided information on the degree of functional relationship between variates. They extracted groups of related characters from the matrix by hierarchial clustering of coefficients at arbitrarily selected levels. General factors such as size were eliminated by use of partial correlation coefficients. Their methodology was criticized severely by Bock (1960) but the model of morphological integration is valuable. Van Valen (1965a) revised the original technique and his version is readily represented in programming packages for numerical taxonomy. Gould and Garwood (1969) showed that morphological integration produced sets of variates similar to those resulting from factor analysis.

No hierarchical analysis of correlation coefficients has appeared for foraminiferal data. Inspection of correlation matrices for mixed data and simple graphical representation of coefficients indicates that variates located on the same structure (chamber, aperture) usually cluster together at high values of r. This is expectable in view of the probable method of chamber morphogenesis (Angell, 1967b) and stability of shape, from chamber to chamber in the taxa investigated. However, I found (1972a) in *Globorotalia miozea* and *G. praemenardii* that variates on two orthogonal projections of a given chamber (one set colinear with the coiling axis, one set normal thereto) tended to show higher correlations within each set than between them. The sample size was small and the influence of a general size factor not investigated but the observation suggests that modifications to one profile of a chamber may occur without great influence on the other.

5. MULTIVARIATE STUDIES OF INTRASAMPLE VARIATION

A form as complex as the foraminiferal shell requires many variates for its quantitative representation. In this respect, biometry is conventionally "reductionist" in its approach to the study of form. An integrated whole is reduced to set of variates for which data are collected. This raises a methodological problem because the shell is an integrated structure and variates need to be considered simultaneously, rather than singly or in pairs. Herein lies the appeal of multivariate methods of analysis. The introduction of these methods to foraminiferal biometry over the last decade is a principal point of departure.

Multivariate data pose a problem of representation when intrasample variation is studied. If m measurements are made on each specimen in a sample, individuals can be envisaged as points in a m-dimensional space. An individual's measurements provide its coordinates in this space. But if $m > 3$ the position of individuals in the space cannot be shown diagrammatically. Principal component analysis (Blackith and Reyment, 1971; Dempster, 1969; Seal, 1966) is a technique for low dimensional representation of points in a multidimensional space. The valuable feature of the analysis is that coordinate axes are produced which provide the closest 1, 2 or 3 dimensional representation of the position of points in the m-dimensional space. Loss of information is minimized. Generally, representation on principal component axes will give a more accurate portrayal of points in the multidimensional space than that provided by any equivalent number of the original variate axes. Plots of individuals on principal component axes permit visual assessment of sample homogeneity. Reyment (1968) analysed seven shell and chamber dimensions in a sample of *Spiroplectammina olokunsui* and found discontinuities indicative of shell polymorphism in the scatter when the data were plotted on principal variates for the largest and second largest principal components. Pronounced discontinuity in the scatter for *Globorotalia pseudobulloides* (Malmgren, 1972) was produced by inclusion of a binary variate (coiling direction) with a set of measurements of chamber and shell dimensions.

The orientation of the principal variates permits their interpretation as either size or shape variates (Jolicoeur and Mosimann, 1960). Direction cosines for a particular axis that are all positive indicate simultaneous increase in all of the original variates. The axis can be identified as representing size variation when it is noted that organic growth can be loosely described as increase in size. To the contrary, an axis with both positive and negative direction cosines reflects shape variation. Increment along the axis is associated with decrease in the original variates bearing negative direction cosines relative to those with positive direction cosines. This represents change in proportions among the original variates. Principal component analyses of characters of the foraminiferal shell have mostly concerned mixed data sets. Typically, the largest component has reflected size variation (Berggren *et al.*, 1967; Reyment, 1961; 1966; 1969; Malmgren, 1972; Scott, 1970) while remaining components reflected shape variation. The variance accounted for by the largest component in these analyses was often well over 50% of the total. This may reflect the broad ontogenetic segment sampled although Scott (1972b) found no appreciable reduction in the proportion absorbed by the first principal component when only specimens with a given number of chambers were sampled. As variates have often been restricted to one per chamber, little has emerged from principal component analyses that is relevant to the study of shape variation in the basic building unit of multicameral shells. However, variations in increment rates between successive chambers have been found to contribute to shape variation (Malmgren, 1972).

The utility of principal component analysis for parsimonious representation of multivariate scatters is governed by the magnitude of covariation among characters. If many characters are highly correlated, as typically occurs in organic structures which increase in size, then a low dimensional representation may be achieved with little loss of information. Note that the representation is no more than that. The configuration may reveal clusters of individuals but further analyses are required to establish objective classifications. Principal component analysis is not a classificatory technique. The rationale of the technique becomes questionable when the original variates are not measured in the same units (Anderson, 1958; Seal, 1966).

As a method for study of multivariate shape variation, principal component analysis is not highly informative. Typically with foraminiferal data, shape variation is partitioned among several small components and there is no facility for synthesizing the information for specified characters. Sprent (1972) remarked that such components are likely to represent a mixture of size and shape information. Moreover, if one component is taken to portray size variation and the others shape variation, then the latter are all independent of the size component. Thus the analysis fails to provide information on how shape varies with size.

Jolicoeur (1963) defined multivariate isometry as equality of direction cosines for the largest principal variate of the logarithmic covariance matrix. This definition is evaluated by Mosimann (1970) and Sprent (1972). The bivariate straight line generalizes to an m-variate straight line if there is only one non-zero principal component. This rarely occurs with foraminiferal data, especially if m is, say, greater than five. In the absence of straight line relationships in multivariate data, the functional relationship approach becomes unsatisfactory and may have to be discarded. Mosimann (1970) in a theoretical study of multivariate allometry, defined isometry as the stochastic independence of a shape vector and a given size variate and showed that shape is isometric in relation to at most one size variate. Choice of the latter thus becomes crucial and dependent on the biological problem investigated. That chosen by Jolicoeur is convenient in that it is invariant under unequal changes of scale.

Factor analytical models and their relation to principal component analysis are clearly described by Seal (1966) who, with Blackith and

Reyment (1971), is critical of their application to biological data. Pitcher (1966) used factor scores to represent relations between fusuline taxa but the techniques discussed in Section E are more appropriate for this purpose. A much fuller account of an application of the same factor analytical model to paleontological data was given by Gould (1967).

6. HOW TO CHARACTERIZE SHAPE?

I have stressed that biometry of the shell is heavily constrained by inadequate data on shape, particularly of chambers. Such is the constraint that, although foraminiferal workers have not as yet researched the problem, it seems justifiable to refer to solutions considered in other fields. Sneath (1967) compared hominoid skulls using coordinates of the substructures. The auditory meati, for example, could be recognized as homologous "the same" in various skulls by their general shape and position. This may be described as a "landmark" approach and will fail when landmarks are not available. Sneath found this applied to the vault of the skull and it is clear that it will also apply to the featureless profiles of many foraminiferal chambers. It is unlikely that this approach will be profitable.

Of much greater applicability are various "scanning" techniques in which the coordinates of the structure under study are systematically recorded. Snee and Andrews (1971) characterized the shape of potatoes by measuring diameters at equal intervals along the major axis (length). This reduces the profile studied to a polygon and variation in polygonal shape was analysed by a split-plot analysis of variance model. Fourier analysis has been applied to shape of sedimentary particles by Ehrlich and Weinberg (1970) and to the ostracode carapace by Kaesler and Waters (1972). Here, radii are measured from the centroid, or reference location, to points on the profile. The angle between successive radii may be constant or varied according to the complexity of the outline. The density of radii determines the resolution of the analysis as the nth harmonic requires $2n$ points for computation. A restriction is that bi- or multivalued data cannot be processed. These arise with embayed profiles, but with the generally convex profiles of foraminiferal chambers this restriction will seldom apply. Amplitude spectra for profiles may be compared graphically or used as data for discriminatory or classificatory analyses.

Vašíček and Jičín (1972) converted continuous curves into equal line segments and represented shape as a sequence of angles made by the line segments with the perpendicular. This approach is allied to procedures in picture analysis wherein boundary lines are encoded as number sequences in which elements specify the direction of unit lengths of the profiles. Freeman (1970) reviewed work in this field. It is clear that the problem faced by biologists in quantifying shape of organic structures is at one with that of the pattern analyst. Intuitive appreciation of shape when man serves as the processor, has to be replaced by a quantitative description of all shape information when a machine is to serve as the processor. As in data capture, significant advances affecting foraminiferal research are likely to be made in quite unrelated fields. Particularly important are theoretical studies of geometries suitable for analysing biological shapes. Blum (1973) focussed attention on the internal symmetry of biological structures rather than on their boundaries. Thus his geometry is based on a point and its growth rather than on the point and line of conventional geometries. It promises a biometry that is revolutionary relative to that reviewed here. Blum is surely at the core of the problem when he asks (p. 282) "if we have no efficient statements for describing biological shapes, how can we find relationships and causes for them?"

E. Mapping Populations: Distance and Discrimination

A primary task in taxonomy, in studies of geographical and temporal variation, and in reconstruction of phylogeny, is objective assessment of affinities between populations. Conversely, granted a biometrical characterization of populations, there is a general requirement for procedures that accurately allocate additional individuals or samples to their respective populations.

1. DISTANCE AND LOW DIMENSIONAL REPRESENTATION

When the data consist of continuous linear measurements, samples can be envisaged as clusters of points in a multidimensional space, as in the previous discussion of intrasample variation. Measurement of inter-sample distance requires, firstly, definition of a reference point within each sample scatter. Almost invariably the point specified by the multivariate mean is taken. It is equivalent to the centre of gravity in mechanics. Secondly, the "route" to be traversed between the samples must be defined. Perhaps intuitively we select the shortest route, that which in the 3-dimensional space of our experience is specified by Pythagoras' theorem (but remember that there are other routes available—the "city-block" route is one). Although there is intrinsic appeal in the euclidean distance defined by Pythagoras, it is essential to note that it is computed using coordinate axes that are at right angles to each other. In a set of multivariate data this implies that the variates are uncorrelated. Seldom is this true of variates on the foraminiferal shell for it is a highly integrated structure. The effect of correlation is to produce excessive measurements of distance. Redundant information is included, proportional to the amount of correlation. The advantage of D, Mahalanobis' distance, (Rao, 1952; Blackith and Reyment, 1971) is that it eliminates redundant information from the pythagorean measure. In principle it is preferable to measures that ignore correlation of variates. However, some empirical studies have shown little difference between Mahalanobis and pythagorean distances (Gower, 1972).

There are other advantages associated with Mahalanobis' distance. As D^2 is computed as the ratio of between sample mean squares to pooled within sample mean squares, it provides under certain assumptions (Reyment, 1962) a test of the significance of the distance between 2 sample means. Rao (1952) described a method for computing D^2 variate by variate and a test of the significance for the additional distance provided by a variate. This allows objective assessment of the importance of particular characters in taxonomic discrimination. Fig. 31 shows the

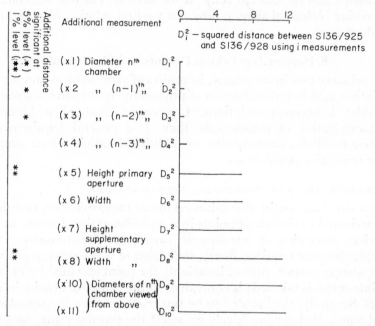

Fig. 31. Increments in D^2 produced by additional characters in comparison of two samples of *Globigerinoides trilobus*. (After Scott, 1970).

increments in D^2 between 2 samples of the *Globigerinoides trilobus* lineage. Two apertural characters are of primary importance. Marcus (1969) described an application of D^2 to the measurement of selection. Burnaby (1966) showed how size information, not necessarily relevant in assessment of taxonomic distances, could be excluded from D^2.

A measure of distance between 2 samples is the simplest of the taxonomist's requirements. Generally, distances among a suite of samples require study. This leads to a matrix of between sample distances and to the problem of low dimensional representation of the samples configured by these distances. A space of $k - 1$ dimensions is required for full representation of the disposition of k samples (supposing that the number of variates exceeds the number of samples).

An important characteristic of a good dimension reducing method is that it provides a low dimensional configuration which, for a given number of dimensions, is the closest approximation to the original high

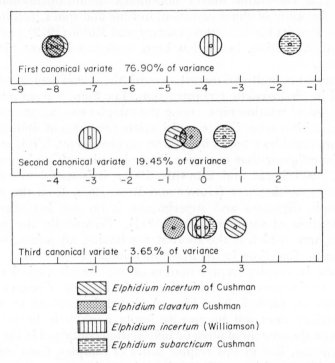

First canonical variate   76.90% of variance

Second canonical variate   19.45% of variance

Third canonical variate   3.65% of variance

Elphidium incertum of Cushman

Elphidium clavatum Cushman

Elphidium incertum (Williamson)

Elphidium subarcticum Cushman

Fig. 32. Configuration exhibited by species of *Elphidium*, characterized by 8 variates, when plotted on 3 canonical axes. The circles indicate 90% confidence intervals. (After Buzas, 1966).

dimensional configuration. The methods usually applied in foraminiferal studies (Buzas, 1966; Scott, 1971) belong, like principal component analysis, to a family of techniques that involve the computation of new variates that are linear combinations of the original ones (Seal, 1966; Gower, 1966). These variates (called "canonical" in statistical literature) are so oriented that they display sample means at maximum distances apart. Geometrically, they may be envisaged as providing a set of best projections. Figure 32 shows a configuration of *Elphidium* species on canonical axes. The utility of the representation is contingent upon a large proportion of intersample variance being represented by a few of the canonical variates.

Another approach to dimension reduction is provided by non-metric multidimensional scaling (Kruskal, 1964a; 1964b), first developed in pyschometry but now widely used in some historical sciences (Hodson *et al.*, 1971). Its merit in historical investigations is that it is order preserving. The analysis attempts to maintain, as far as possible in the low dimensional space, the order of items in the original high dimensional space. This constraint makes non-metric multidimensional scaling suitable for study of clinal variation, in time and space. Jardine (1971) demonstrated its use in anthropometry and Rohlf (1972) provided an empirical evaluation, but it has been neglected in most biometrical disciplines.

Assessment of multivariate distances within a set of geological samples is useful in phylogenetic reconstruction. The known geographical and stratigraphical relationships among the samples can be compared with their disposition according to multivariate measures of similarity. This procedure provides a test of hypotheses on phylogeny. If it is acceptable to the investigator that evolution is parsimonious (Camin and Sokal, 1965; Cavalli-Sforza and Edwards, 1967), it is useful to compute a minimum spanning tree (Gower and Ross, 1969) from the matrix of intersample distances and superimpose it on the low dimensional representation of samples (Scott, 1971). Technically the minimum spanning tree, which is the shortest net linking all points, will draw attention to distortions in the low dimensional plot of sample positions due to the incomplete representation of intersample variation. But to the taxonomist there is special interest in the location of branches which might indicate phyletic branching. A branch in a minimum spanning tree certainly does not necessarily imply a phyletic branch. Much depends on the content of phyletic information contained in the distance measurements. But in this respect suites of samples, from established stratigraphical positions and covering a large time span, place the foraminiferal worker in a position much superior to most of his biological colleagues. As yet, this intrinsic advantage has been neglected.

Analysis and representation of distances in multidimensional spaces has developed rapidly in the last decade. The review presented relates primarily to configurations of samples not individuals. This is intentional, in view of the merits of sample-based investigations of populations argued later. But it should be noted that representational techniques applied to sets of individuals, configured by a matrix of similarity coefficients can be applied to sets of samples configured by a matrix of distances and vice versa (Gower, 1966; 1972). The hierarchical representation (dendrogram), almost universally used in the assessment of relations between individuals in numerical taxonomy (Sokal and Sneath, 1963), can be usefully applied to a matrix of between sample distances if there are grounds for suspecting hierarchical relationships. Conversely, matrices of similarity coefficients representing affinities between individuals can be analysed by canonical variates, originally applied to sample-based measures of distance.

2. DECISION MAKING IN TAXONOMY

In multivariate statistics, discrimination refers to rules for allocating new material among known populations. Rules that minimize the risk of misallocation are sought. One that achieves this for multivariate normal populations with common variances and covariances, is given by Fisher's linear discriminant function (Rao, 1952). A linear function of the variates is used to define a hyperplane so orientated that samples are at maximum distances apart. In this orientation the allocation boundary passes through mid-point of the line joining the means. The variance of the discriminant scores is Mahalanobis' D^2. This statistic provides a test of the amount of overlap of the samples and a means of estimating the error in allocation (Berggren et al., 1967; Scott, 1966a). The intimate relation between Mahalanobis' distance and statistical discrimination further supports the wide use of his measure in biometry.

At first sight, discriminant functions may appear to solve the problems of identification that beset taxonomists and obviate much subjective decision making. This is not so. The basic disadvantage is that the technique assumes that the specimen presented for allocation is drawn from one of the populations represented in computations of the discriminant function. There is no provision for a decision of the type "not a member either of population A or B". Thus the investigator must accept, prior to determining the score for a particular specimen, that it comes from one of the populations A, B. To the practising taxonomist this restriction entirely overrides the advantages of the technique. It also emphasizes that discriminant function analysis is an allocation procedure, not a classificatory procedure. Very seldom in foraminiferal taxonomy are these operations entirely divorced in routine research. Classifications cannot usually be accepted a priori.

Another restriction concerns the set of characters studied. It must be the same in both samples and, preferably, each variate should be measured in the same units. Obviously, any character always absent from one population but present in the other produces complete discrimination. This may not correspond with taxonomic distinctions. For example, supplementary apertures in Globigerinoides are present only on the latest chambers formed in ontogeny. Intrapopulation discrimination on the basis of this structure would have no taxonomic validity. A set of mixed variates (some linear measurements, some qualitative) cannot be processed satisfactorily by discriminant functions (Kendall and Stuart, 1966) and the analysis fails when the number of variates exceeds the sample size. Dempster (1969) discussed ways to circumvent this problem although, with foraminiferal data, it may seldom arise. The space in which the discriminating surface is constructed is defined by the pooled sample covariance matrix and it is assumed that individual covariance matrices are homogeneous. In my experience with foraminiferal data this is exceptional. Reyment (1962; 1969) discussed reasons for non-homogeneity. Allocation errors may increase when heterogeneity occurs.

Clearly, discriminant functions could serve a useful role in taxonomic practice for groups where stable classifications already exist. Standards of data collection will need to rise substantially if an allocation rule worked out by one taxonomist is to be generally applied by others. A counsel of perfection by present day standards, but certainly needed in the future, is that operator error estimates should accompany any discriminant function proposed for wide usage. Discriminant function models of greater generality, especially those that dispense with assumptions about population parameters, are required. Currently, the discriminant function in foraminiferal research is a toy for the biometrically inclined rather than a practical aid in taxonomy.

F. Classification

Classification is the ordering of organisms into groups (Simpson, 1961). As such, it is a fundamental activity (Pratt, 1972b) logically prior to discrimination, the allocation of individuals or samples to predetermined taxa. The question, "is this sample drawn from one or more populations?" is a classificatory question. Tan's analysis (1932) of the frequency distribution of the number of nepionic chambers in samples of *Cycloclypeus* is an example of subjective extraction of classificatory information from numerical data. More satisfactory methods for resolution of populations that are represented in heterogeneous samples were described by Cassie (1954), Harris (1968) and Mundry (1972 and references therein) for univariate normal data. Day (1969) described a procedure for multivariate normal data in which components are drawn from populations with equal covariance matrices.

These and similar techniques extract classificatory information from the form of the sample distribution. Multivariate methods of this type usually operate on data in a space specified by the sample covariance matrix. Classificatory methods based on sample distributions should be clearly distinguished from exemplar methods in which the variance–covariance structure of the population is neglected. Numerical taxonomy (Sokal and Sneath, 1963; Jardine and Sibson, 1971) is almost exclusively concerned with exemplar methods in which all classificatory information is derived from single individuals.

Sample-based classificatory techniques are at an early stage of development. Indeed, my previous criticism that foraminiferal biometricians have failed to classify *de novo* may be rebutted by the contention that adequate techniques are not yet available. At the specific and intraspecific levels, there is a broad requirement for sample-based methods of classification capable of operating on data sets containing several types of observation (linear measurements, counts, attributes) and which make a minimum of assumptions about either the number or the form of the populations that may be represented in the sample.

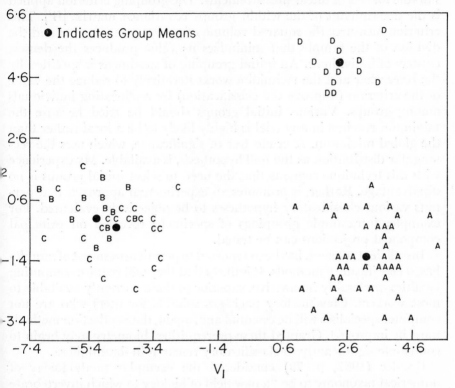

Fig. 33. Representation on canonical axes of classification of nummulite specimens using successive two-group divisions found by projections of multivariate data from the best discriminating surfaces. In qualitative taxonomy, groups A, D are from distinct species but B and C are from the same species. (After Wright and Switzer, 1971).

These specifications, although still insufficiently general, have yet to be met by mathematical statisticians. Current methodology is exemplified, firstly, by a classificatory procedure applied to Eocene nummulitids by Wright and Switzer (1971). Discriminant planes at random angles are computed from the sample covariance matrix (any *a priori* knowledge of specimen groupings is neglected). The projection that provides the best two-group separation of individuals, using as a criterion the ratio of between groups variance to within groups variance, is selected for classificatory purposes. Once found, a group may be subjected to the same procedure to obtain further splits. The final groupings (Fig. 33) is examined by discriminant function techniques to assess the significance of intergroup distances.

Marriott (1971) described a sample-based classificatory technique that is a development of work by Friedman and Rubin (1967). It is suitable for sets of linear measurements. The grouping criterion applied is the determinant of the within groups covariance matrix, $|W|$. This criterion measures the squared volume of pooled group spaces and the division of the sample that minimizes its value produces the densest clusters of individuals. An initial grouping of specimens is specified by the researcher and the technique works iteratively to reduce the value of the criterion (improve the classification) by reallocating individuals among groups. Various initial groups should be tried because the minimum reached in any trial is highly likely to be a local rather than the global minimum. A crude test of significance, which uses the rectangular distribution as the null hypothesis, is available. My experience with this technique suggests that the need to select initial groups is no disadvantage. Rather, it promotes an experimental approach that permits various classificatory hypotheses to be objectively evaluated. For example, intrasample groupings of specimens revealed on principal component projections can be tested.

Insufficient research has been reported to permit assessment of sample-based classificatory methods. It is likely that they will require computing facilities, preferably interactive, superior to those currently available to most workers. Classificatory packages suitable for users who are not computer specialists will be essential and, again, data collection methods must be improved. Granted these prerequisites, biometry may begin to contribute significantly to classificatory research in foraminifera.

Kaesler (1967, p. 79) considered the exemplar methodology of numerical taxonomy to be "a new field of biology to which invertebrate paleontologists can make significant contributions, a field calling out for new ideas developed from experience with the fossil record and geologic time". He urged workers to respond to the call. The call has not been answered by foraminiferal taxonomists. Perhaps this is fortunate. Intrapopulation variation is vital information in taxonomic and evolutionary studies and in interpretative investigations of shell form. Any methodology that ignores intrapopulation variability can make but small contributions to these studies.

Some aspects of current progress in numerical taxonomy were demonstrated by Kaesler (1970) with fusuline data. The analysis dealt with five specific or subspecific taxa described by Dunbar and Skinner (1937). These taxa were accepted as primary data or operational taxonomic units (OTU). No justification for accepting the results of prior, qualitative, classificatory research was given. Sokal and Sneath (1963, p. 121) dealt with this point by stating that, "although taxonomists make objective tests of the criteria for specific distinction in only a small minority of cases, their judgement in this matter generally inspires confidence (certain difficult taxa and indifferent taxonomists excepted)".

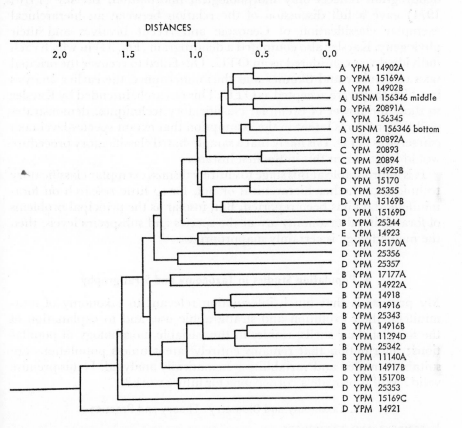

Fig. 34. Dendrogram produced by unweighted pair-group clustering with arithmetic averages from distance matrix, 38 characters, for 34 specimens of *Pseudoschwagerina*. Clusters show weak relation with the classification (taxa A–D) employed by Dunbar and Skinner (1937). (From Kaesler, 1970).

This is rationalization. Mean values of 63 characters were used to represent each OTU. Kaesler differentiated this procedure from the single specimen exemplar method but it is only a close variant. In both, information on covariance structure of the populations represented is excluded. Thus the distance matrix computed to show similarities among the OTU made no provisions for correlation between characters and redundancy of information. Blackith and Reyment (1971) discussed the distinction between the variate spaces of exemplar and sample-based techniques. The dendrogram (see Jardine and Sibson (1971) for recently developed cluster methods) for the fusuline OTU corresponded rather poorly with their phylogenetic interpretation by Dunbar and Skinner. Admittedly, there is no necessary reason for close correspondence as the dendrogram reflects only morphological information. Bretsky (1970; 1971) gave a full discussion of the relation between an hierarchical exemplar classification of Cenozoic and recent bivalves and their phylogeny. Kaesler also computed a dendrogram (Fig. 34) in which each individual was considered as an OTU. This failed to recover the original taxa of Dunbar and Skinner and thus undermined the earlier analysis in which they were accepted as OTU. This research, intended by Kaesler to show the power of exemplar classificatory techniques, demonstrates well the problems latent in the assumption that extant species-level taxa can serve as OTU. It is likely that a sample-based classificatory procedure would have shown this in the first instance.

It is unprofitable at this stage to entirely dismiss exemplar classificatory techniques, whether hierarchical or not, for so little research on foraminiferal data has been reported. But, insofar as the principal problems of foraminiferal taxonomy are at the species and subspecies levels, then the methodology is certainly inappropriate.

V. Population Studies in Taxonomy and Stratigraphy

My premise is that much information relevant to taxonomy of foraminifera, their evolution and stratigraphic use, and to explanation of the form of their shells, resides in the variable morphology of populations; information that remains entirely latent unless populations are suitably sampled and variability systematically analysed. Is this premise valid? Will it provide a cornerstone for future work?

A. Taxonomy and Biometry: Theory and Practice

1. SAMPLES AND EXEMPLARS

The premise that the peculiar contribution of biometry to foraminiferal taxonomy is the provision of organized information on population variability can be dismissed as irrelevant if taxonomic data in fact reside in particular individuals, not in the population at large. There is a long history of practice that suggests that the premise has not been acceptable. In the last century F. W. Millett (quoted in Newell, 1947, p. 163) urged students to be "very careful to select typical specimens only, the intermediate things are a nuisance and should be severely ignored, unless the types are absent from the deposit". This might be dismissed as an example of a typological stance that disappeared with the revolution in systematics that occurred in mid-twentieth century, but is not modern work on planktonic foraminifera substantially similar? Brönnimann and Resig (1971) considered that population studies were unrealistic for fossil material because samples may contain individuals from many death assemblages. How this constraint justifies their adherence "as closely as possible to the typological approach" (p. 1267) is not explained. Blow (1969) investigated 228 planktonic species from late Paleogene to Quaternary strata and found from a "re-examination of deposited types" (p. 312) that only a few required description as new. Repeatedly in this work, emphasis was placed on the morphology of particular specimens. For example, *Globigerina calida calida* was restricted (p. 317) to "the extreme morphological type exemplified by the holotype and the paratype (especially) figured by Parker". Blow's concept of *Globorotalia margaritae* (p. 363) was based on observations of the holotype and a number of metatypes. These types were all found to be in complete morphological agreement. In the same vein Banner and Blow (1965, p. 112) wrote that "selection of lectotypes is necessary in order to establish firmly the precise morphology of a taxon". The primacy of the type specimen over all others as a source of taxonomic information is also suggested by taxonomic descriptions (Blow, 1969; Jenkins, 1967a) that consist of a minute account of the morphology of the holotype only. These authors do not deny that foraminiferal taxa are variable. Rather, their (implicit) argument is that characters essential to the definition of a taxon occur in a few, perhaps only one, specimen.

If taxonomic information does reside in types then the role of biometry in taxonomic research is remarkably circumscribed relative to the role it should play if taxonomic information is to be extracted from populations. Typology implies that the relevant material of the biometrician is the set of type specimens. Studies of population differentiation that take into account character covariation, emphasized in the previous chapter, become irrelevant because classificatory data does not reside in the population. In fact, typology implies that sample-based methods of analysis are redundant and should be replaced by appropriate exemplar techniques of numerical taxonomy. Sample-based methods might not even be relevant to the task of selecting the type specimen. In principle, no two members of the set of type specimens have identical characters. Measures of central location, variance and covariance are not necessarily appropriate for selection of an individual fulfilling these requirements.

In contrast with the typological approach to taxonomy, the theses of the "new systematics" (Huxley, 1940) emphasized the importance of population variability and the need for systematic analysis of variation. In delineating the future of taxonomic research Huxley wrote (p. 39) "First and foremost comes the need for more measurement". Taxonomic studies, wrote Simpson (1961, p. 65) "are always statistical". Populations, not individuals, are classified thus sample-based techniques are appropriate. Moreover, the theses of the "new systematics" and the empirical data on which they were based strongly suggest that taxonomic research requires multivariate analyses of variability. The inheritable information of a species is carried by members of its contemporary populations. Only in special cases is the information held by two individuals identical. Even in these cases, replicate genetical information does not necessarily lead to replicate morphology. A character may be widely distributed in the population but is not necessarily universal. Beckner (1959) gave a set-theoretic interpretation of these data and showed that a species is polytypic relative to its set of taxonomic characters. Each individual possesses a large number of characters in common and each character is possessed by a large number of individuals. No one character is sufficient for inclusion of a particular specimen in a given species. The latter statement cannot be strictly proved but it is plausible. Beckner shows this formally by arguing that if, in fact, all individuals of a species did possess a common character which was not found in any other species, then the subsequent discovery of this character in an individual of a second species would be insufficient evidence for its transfer to the first.

The foundation I have suggested for biometrical studies of foraminifera is taken from the "new systematics". Kaesler (1967), proselytizing for numerical taxonomy, declared that the species concept of the "new systematics" was non-operational. His supporting argument centred on the inapplicability of the interbreeding criterion to fossil populations. But this argument will not do. Huxley wrote (1940, p. 11) "there is no single criterion of species . . . a combination of criteria is needed". Kaesler has overlooked the polytypic nature of the new systematists' concept of species, whether of the category itself or of taxa of species rank. The theses of the "new systematics" cannot be rejected on grounds that the interbreeding criterion is not an operational procedure for fossil organisms. For it was not held as a necessary criterion for recognition of species. "Operationalism" calls for the use of practical, objective, repeatable procedures in taxonomy, and suppression of intuitive practices. In no way are the techniques for analysis of population variability, described previously, less operational than those dealing with exemplars. Indeed, a basic inconsistency in numerical taxonomy is that it stresses (quite justifiably) that classificatory techniques must be objective but offers no operational procedure for recognition of OTU or selection of exemplars. The intuitive element is not eliminated. The failure to classify at the basic population level is implicit in Pratt's criticism (1972b) of numerical taxonomy. Typological practices, whether in conventional qualitative description of taxa or in numerical taxonomy, fail to analyse polytypic populations. This is why they must be rejected and the cornerstone of biometry be that of the "new systematics".

2. QUALITATIVE STUDIES OF VARIATION

Rejection of typology and acceptance of a population approach by some taxonomists has not necessarily led, in turn, to employment of biometry. Wade (1964) accepted enthusiastically the implications of modern evolutionary theory for taxonomy and the need to establish, clearly, the natural morphological limits of biological species, but did not think that "large numbers of specimens and painstaking measurements" (p. 276) were essential to delineate lineages. Hay (1971, pp. 126–7) argued that "verbal description of the degree of variation of size, shape, and texture characteristic for a particular population is virtually impossible, but illustration of suites of specimens of a particular species can adequately transfer the necessary information to an independent researcher". Such preference for pictures over words may serve for qualitative description; indeed the philosophy appears to be one in which the user is provided with good visual images and left to formulate his own description. However, I question whether visual comparison of suites of good illustrations leads to objective analytical results in the classification and discrimination of populations.

Cole recognized that many larger foraminifera are highly variable but rejected quantification and statistical analysis. He wrote (1964) of Drooger's work, that species were recognized by mean values that did not reflect intra-sample variation in basic growth pattern. Undue splitting was seen as another consequence of the application of biometry to populations studies (Cole, 1963). While I accept his contention (1963) that analysis of variation by statistical methods may easily lead to confusion, not enlightenment, his preferred method of qualitative assessment of comparative series of individuals is not free from the same charge. Consider Cole's analysis (1966) of variation in a nummulitid species, *Camerina panamensis*. Specimens from a single locality were arranged in a series that reflected their similarity of form. Criteria used to judge similarity were not elaborated; neither was sample size nor sampling technique. The analysis then proceeded to judgement of morphological similarity among selected figured specimens. Again, no information about method of selection was given. Cole stated that some are larger, some thicker, some more tightly coiled than others. But (p. 240) these are "relative and gradational features from specimen to specimen". All "have the same fundamental arrangement of the chambers and the same structures". These constitute the entire data. His conclusion is that all the specimens are essentially the same. Only one species is recognized because Cole (p. 241) "cannot conceive of gametes from specimens so similar to each other repelling one another". Excellent sections of the selected specimens show chamber shape and shell geometry in detail but the history of shape variation in the ontogeny of the shells and its possible explanatory use are entirely neglected. This work in no way establishes that there is continuous variation within the sample. The figures support only the contention that variations in form occur. The reader remains unenlightened about the extent of intergradation of specimens at the locality.

Cole's many researches on larger foraminifera are remarkable for their insistence that species are variable. Yet his qualitative accounts do not implement the philosophy of the "new systematics". Research into polytypic taxa calls for mapping of character distributions within and between populations, a search for characters in common and for the location of discontinuities in distributions. Such is the capacity of the human mind that these tasks require quantitative methodology if they are to be performed objectively.

3. THE PROSPECTS FOR A QUANTITATIVE TAXONOMY

Progress towards a quantitative taxonomy is exemplified by Meulenkamp's research (1969) on uvigerinid taxa. This work is as advanced as any in its practical attempt to quantify form and apply information on variability to the definition of taxa. *Uvigerina cretensis* (p. 141) is defined as "A *Uvigerina* of the *U. cretensis* group with a mean number of uniserial chambers higher than 2.10 and $\bar{s}_2$ values higher than 75". There follows qualitative descriptions of chamber arrangement, aperture position and shape, and ornamentation of chamber walls. Mean values of eight variates, some with error estimates, are given for the type sample. Diagnostic information is mainly qualitative.

The study is founded on analyses of populations. This is typical of almost all biometrical studies of foraminifera. It is a good augury. If there is a revolution required in the application of biometry to foraminiferal taxonomy then it involves the superstructure rather than the foundations. These interlock with those of the "new systematics".

The inadequacies of modern applications of biometry to taxonomic studies are also mirrored in Meulenkamp's research. Fifteen variates were considered but some were ratios of others. Variates related either to gross dimensions of the shell or to types of chambers (enumerative data). These are the data that can be objectively recorded from the foraminiferal shell with the laboratory equipment normally available. Shapes of chambers, apertures and ornament are neglected, I think, not because the biometrician considers them irrelevant, but because he cannot readily and reliably quantify them. Thus the primary inadequacy of biometrical practice in taxonomy is one of data capture. Here the prospect is bright. In the last three decades there have been major advances in microscopy, automatic methods of recording data and in computation. Equipment required to quantify characters of the shell that are presently neglected does exist. In the next decade suitable data logging and analytical systems should resolve the problem long faced by foraminiferal workers.

The second inadequacy of biometry in taxonomic research is analysis. Biometry has not penetrated into the classificatory process. Rather it is used to characterize certain aspects of morphology and to assist in allocation of specimens. Meulenkamp did not apply a quantitative classificatory procedure to his data to show that the taxa he recognized are optimally located relative to character distributions. Now, I agree that such evidence is insufficient for taxonomic proposals but it would provide a substantial justification. To a degree, the analytical failure seems to be contingent on inability to quantify much of the morphology of the shell. Information on chamber shape, for example, is essential at the specific level of classification and until it is available "words and pictures" will prevail. The analytical failure is also contingent on the development of geometries suitable for studies of biological shape and on progress in theoretical statistics. Very general, non-parametric methods for classification of multivariate populations have still to be developed. Finally, foraminiferal biometricians have tended not to extract the maximum amount of taxonomic information from their data. A general failure with ontogenetic data is apparent; relative growth studies have been neglected. Here, better education in biometry is required. When these deficiencies have been remedied, and it is likely that the next decade will see substantial progress, biometry will become an integral part of taxonomic research, not an occasional supplement.

B. Biometry, Stratigraphy and Evolution

1. STRATIGRAPHY FROM EVOLUTION

Historical sequences of fossils preserved in rocks provide primary data for the study of evolution, its pathways and processes. In modern evolutionary theory, population studies occupy a central place in the methodology required to map the history of lineages and the loci of divergences. No less is the population concept central in the interpretation of the patterns so delineated. This outlook is probably acceptable to most foraminiferal workers but it has not served to orientate their research. Van Valen (1965b, p. 188) recalled that over half of the world's paleontologists study foraminifera but "despite material scarcely paralleled elsewhere in abundance, little of value for the general study of evolution has appeared from it until now". The reason why evolutionary studies have been neglected is clear. Many foraminiferal workers are engaged in stratigraphical investigations. In stratigraphy there is a demand for criteria useful in correlating strata. Suitable criteria include trends in morphology of fossils that are unidirectional, vertically, from bed to bed and that are devoid of obvious local facies influences. Strata containing populations at a common stage of advance on such a character trend are regarded as correlated. Most workers attach an implication of identity in time to the term "correlate", but in terms of what is falsifiable by data of this type, only identify in order of events is established.

The essential point is that the search for correlative criteria has led to a partial investigation of evolutionary patterns. Consistent vertical changes in morphology (chronoclines) have been sought. Geographic variation tends to be ignored as the aim is to provide correlative data over the widest area possible. Characters are considered singly because of resulting simplicity in determining identity of samples or degree of advance. Thus certain "straight" limbs of evolutionary trees have been mapped in terms of selected, unitary, characters. Research is directed towards consistent results in stratigraphy, not on integrated studies of evolutionary pattern. Biometricians have reflected this outlook. Instead of "stratigraphy and evolution" we have "stratigraphy from evolution".

2. STRATIGRAPHICAL DATA FROM MORPHOLOGICAL TRENDS

The incentive for biometrical investigation of morphological trends is the possibility of greater stratigraphic precision than that obtainable from qualitative assessment. However, the fundamental problem for quantitative and qualitative workers alike, is identification of lineages. Stratigraphical results rest on the assumption that the populations sampled represent the same lineage. Kennett (1966) mapped a progressive change in the conical angle of the final chamber of specimens in the *Globorotalia crassiformis* bioseries. As the conical angle increased, the axial profile of the chamber became rounded and the peripheral keel typical of early populations tended to disappear. Kennett's concept of the lineage thus included forms with, and without, a keel. Blow (1969) rejected this concept. Forms with a peripheral keel, he considered, are best developed in warm water masses, non-keeled forms in cooler water environments. On this view, the trend in conical angle reflects the intermingling, in changing proportions, of populations of two lineages that Kennett had failed to distinguish. Blow's interpretation is contentious but from it two points arise. The history of the lineage and its relation to its contemporaries must be well established before reliable stratigraphical information can be extracted from it. Secondly, it draws attention to the possibility that a unidirectional trend can be produced by non-evolutionary processes.

Thus far, biometricians have relied heavily on previously established stratigraphical correlations for the detection of parallel and convergent trends. Is this approach adequate? Freudenthal (1969) established that, in Mediterranean Neogene strata, there was progressive and gradual reduction in the number of nepionic chambers in populations of *Planor-*

bulinella. Good stratigraphical control by superposition of strata was available, and he concluded that the populations belonged to one lineage of which the recent species *P. larvata* was the most advanced member. The number of nepionic chambers in *Planorbulinella* from Neogene strata in other regions was also examined. Freudenthal found that *P. zelandica* (New Zealand) was more advanced on the trend to reduction of nepionic chambers than any of the Mediterranean upper Miocene populations examined. If populations from the Mediterranean and New Zealand belong to the same lineage, the stratigraphical implication of this result is that New Zealand strata with *P. zelandica* are younger than upper Miocene. Freudenthal did not accept this conclusion because fossils associated with *P. zelandica* indicated correlation with upper Burdigalian to Helvetian rocks in the Mediterranean region. The argument that *P. zelandica* is not a member of the Mediterranean lineage is strengthened by the observation that the mean number of nepionic chambers in *P. zelandica* is even less than in the recent Mediterranean *P. larvata.*

Freudenthal's study provides an example in which homoplasy (resemblance not due to inheritance from a common ancestor) was detected because of the substantial stratigraphical separation of some of the populations (Recent:Lower Miocene). Where the separation is smaller and the evidence indicating separation is weak, or in conflict, the possibility increases that data from several lineages are unwittingly compared. Data assembled by O'Herne and van der Vlerk (1971) on quantitative trends in several larger foraminifera from two Indonesian localities (Larat and Bg 312) exhibit some aspects of this problem. Values for a measure of the symmetry of the periembryonic spirals in *Miogypsina* indicated that the populations were at a similar stage of evolution. This suggested that the strata could be correlated. To the contrary, a variate expressing the angular relation between the embryonic chambers and the apicalfrontal line of the shell in the same miogypsinids, indicated that Larat was stratigraphically superior to Bg 312. The authors' conclusion that, in fact, the locality at Larat was stratigraphically inferior to Bg 312 was supported by the grade of enclosure of the protoconch by the deuteroconch in *Lepidocyclina* and by the number of nepionic chambers in *Cycloclypeus*. This order was also indicated by planktonic species occurring at the localities. The preponderance of evidence was accepted and that from *Miogypsina* discounted. But what would have been their conclusion if only that genus had been present at the two localities? O'Herne and van der Vlerk did not investigate the reasons for the discrepant data from *Miogypsina* trends but confusion of lineages is one explanation requiring study.

Certainly, stratigraphical correlations established by other fossils may provide a good test for homoplasy, as in Freudenthal's research, but the test is not a sufficient test. Congruence of a character trend with other stratigraphical data does not prove that the populations considered constitute a lineage. If the data selected for the congruence test are false, then the procedure totally fails. It is easy to envisage a circular situation wherein homoplasy is undetected because the test criteria included undetected homoplasy. I suggest that the methodology of future research into extraction of stratigraphical data from lineages should be redirected. The primary requirement is for multivariate mapping of populations in place of the univariate studies that constitute the present literature. So far, selected characters have been studied in isolation. Obviously, the characters selected are those that exhibit strongest chronoclinal variation. In turn, such characters reflect the relative strength of selective pressures in the organism: environment interaction. For a given basic mode of shell organization, common trends in morphology to improve adaptation are likely to arise in lineages that are but distantly related. Cifelli (1969) has sketched this pattern of evolution in the planktonic foraminifera. Briefly then, the character selected as a source of stratigraphical information is the very character that is likely to have appeared iteratively throughout the history of the higher taxon considered. Reduction in the number of chambers per revolution in biserial, triserial and multiserial taxa provides an example. This trend has been quantified by Grabert (1959), Meulenkamp (1969) and Scott (1965). The rationale of a multivariate study is that "conservative" characters, those that for various reasons are not directly involved in optimizing the adaptation of shell form, will be included in the mapping and the probability of detecting homoplasy increased. Hottinger's argument (1960) that biometry is useless for population discrimination among homoplastic lineages because similar ranges of variation, for a specified character, may occur in different lineages (Fig. 35), is cogent only when a single variate is considered. Even a bivariate study may clarify lineage concepts. A trend in number of adauxiliary chambers in *Lepidorbitoides* suggested to Mac-Gillavry (1959) that the samples he examined were from a single lineage. But when increment in size of embryonic chambers was also considered, the samples fell into two groups with the trends in different combinations. Techniques for low dimensional representation of multivariate samples are well known. A change in the outlook of foraminiferal biometricians is the principal ingredient required.

3. EVOLUTIONARY INTERPRETATION

Biometricians have abstracted data from evolutionary processes to provide insights into stratigraphy; but the converse, wherein the stratigraphical record of lineages is searched for insights into evolutionary processes, has received little attention. Because of their stratigraphical value, unidirectional trends in shell morphology have attracted most

comment, although it is not established that they are even a principal
feature of foraminiferal evolution. The apparently non-adaptive nature
of trends has been mentioned repeatedly.

Tan (1932, p. 110) interpreted the progressive reduction of nepionic
chambers in the cycloclypei as an example of orthogenesis, "in which a
direction of specialization once set in, is continued independently of

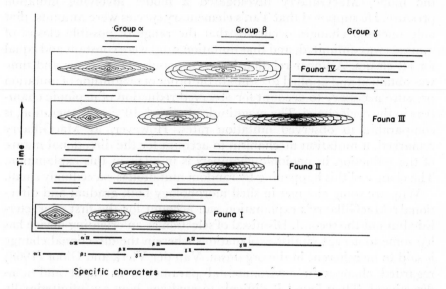

Fig. 35. Simplified example, from studies of shape in alveoline lineages, to demonstrate
that populations from separate lineages may exhibit similar ranges of variation
(horizontal bars at base) for a given character. Thus variation in population IV of
lineage β is closely coincident with that in population II of lineage γ. The character
used in the example was the ratio between axial and spiral diameters of the shell. This
provides limited characterization of shape and does not, for example, reflect the blunter
axial extremities of the specimen from β IV relative to the individual from γ II. From
this scheme Hottinger (1960) argued the inutility of biometry in lineage discrimination
but it is unlikely that this conclusion would have been reached if more variates,
providing better characterization of shape, had been employed. (After Hottinger,
1960).

changes of environment, sometimes even giving rise to overspecializa-
tion". Tan considered that an orthogenetic trend progressed saltatively.
This view arose from the peculiar character of his frequency distributions
for nepionic septae, already discussed. Insofar as these distributions have
not been corroborated by later research, the thesis is currently unsup-
ported. However, the orthogenetic character of the trend leading to
reduction of nepionic chambers has not been rejected by some modern
workers. MacGillavry (1956) believed that it was unlikely that the
trend could be a response to selective pressures because it appeared to

be independent of facies influences and proceeded at different rates in different lineages. The co-existence of ancestral *Heterostegina* (no reduction in nepionic chambers) with cycloclypei was used as an argument against adaptive advantage accruing from reduction in numbers of nepionic chambers. To explain the very gradual, but directed, nature of the change and the fact that it affects the range of variation as well as the mode, MacGillavry investigated a model involving mutation pressure. He supposed that Tan's elementary species were mutants, that only one-step changes occurred, that the range of possible classes of mutants was unlimited, and that mutation rates were constant and equal for all classes. A linear relation between population average and time was obtained and applied to estimate the order of magnitude of mutation pressure necessary to account for observed reduction of nepionic chambers in the cycloclypei. The magnitude obtained, 10^{-7} per generation, is comparable to observed mutation rates. However, as MacGillavry remarked, a mutation mechanism to account for the directional nature of the reduction in nepionic chambers is implicit in this explanation. The chance of this happening by random mutation is exceedingly small.

Why are some changes in shell morphology both gradual and directional? MacGillavry's explanation can account for the first characteristic but not the second. Dismissal of selection as an orienting device has led some to accept vitalist explanations wherein the directional change is said to be inherent in the organism. Van der Vlerk and Gloor (1968) regarded changes in environmental parameters as rapid and non-directional. They found it difficult to envisage how an infinitesimally small change per generation, much less than the tolerance range of the organism, could have selective advantage. From their knowledge of evolutionary trends in lepidocyclinids they concluded (p. 63) that genotype "not only opens evolutionary possibilities but also may channel some of them in an irrevocable way and irrespective of environment". Hofker (1968) viewed increment in size of gavelinellid taxa as an orthogenetic trend which, in turn, necessitated increase in pore dimensions. When the latter trend became unfavorable the problem was sometimes solved (p. 20) "purposefully by the organism itself" by divergence of a new species with smaller pores. Of such vitalist arguments Comfort (1972) remarked that they are without scientific and practical importance. Is their recurrence in the foraminiferal literature a measure of its distance from current biological frontiers?

More conventionally, Bettenstaedt (1958; 1962) interpreted biometrical data on unidirectional trends as demonstrating directed selection (orthoselection, see Rensch, 1959; Hofker, 1959). Bettenstaedt's work is representative of the present contribution of biometry to evolutionary studies of foraminifera. It is limited to empirical data on univariate trends. Interpretation of trends in terms of adaptive advantage is qualitative, and quantitative evaluation of competing hypotheses is eschewed. Orthoselection is used to account for the data but the data are not used to account for orthoselection.

Mechanical efficiency of shell structures relative to their functions is a critical element in the interpretation of unidirectional trends. If a trend is adaptive then improvement in efficiency during its course is expectable whereas improvement is not a necessary component in a non-adaptive trend. Admittedly, as Rudwick (1964) observed, non-adaptedness may be difficult to positively demonstrate with fossils. To do so requires complete knowledge of function. For the shells of foraminifera this is rudimentary. But a feasible step is to evaluate the efficiency of structural trends in relation to possible functions, as advocated in the next section. Decisive evidence may not be produced but the data should at least suggest the extent to which trends are adaptive.

C. Shell Design

Explanation has a primary role in science. Foraminifera can be classified and their evolutionary history disentangled and applied to stratigraphy, but these tasks remain technical insofar as explanatory hypotheses about shell architecture are not generated. Yet the value of research into shell design has often been questioned or discounted. Smout (1954, p. 15) wrote that "the final form of the test is of little biological importance . . . the diversity of morphology within this superfamily (the Rotaliidea) is itself a proof of the slight importance of the final shape of the test . . . mechanical strength is not important". McGowran (1971) evaluated foraminiferal taxonomy and found that ignorance of function and of the process of functional adaptation was not significant. He considered the evolutionary pattern of shell structures to be fundamental in taxonomy. Loeblich and Tappan's classification of the order (1964) used characters that were selected without regard to their function, known or hypothesized. Again, non-selectionist interpretations of evolutionary trends have not led to investigation and explanation of shell morphology.

I suggest that these attitudes should not serve as antecedents for future work, that knowledge of the function of structures is valuable in their taxonomic application and vital in interpretation of the evolutionary patterns exhibited by foraminifera. For investigation of shell architecture, the synthetic theory again offers a suitable model. Evolution by natural selection suggests that the mechanical design of shell structures, relative to their function, is constantly under pressure in the organism: environment interaction. The theory thus justifies analysis of shell structures to evaluate their efficiency for known functions or for a set of competing, postulated, functions. With fossil shells it is impossible to demonstrate that a structure did perform a particular function. But the approach does permit alternative designs to be tested for efficiency in performing such a function. It is possible to demonstrate that a structure *could* function in a particular way (Rudwick, 1968).

Methodology appropriate for functional analysis of fossils has been set out by Rudwick (1964) and its quantitative aspects elaborated by Gould (1970). The roles of biometry in this research are twofold. It can provide basic information on shape, and changes in shape, both during ontogeny and along the morphological pathway traced by the lineage. At the intrapopulation level, relative growth studies can be used to investigate the relation between size and shape and may suggest hypotheses about size-required changes in shape. The question "why this isometry" may arise as well as the more general question "why this allometry". Research on the patterns of coefficients of variation and of correlation among variates may delineate clusters of structures that were linked functionally. Interpopulation studies permit the history of structures to be traced. Discriminatory analyses serve to identify features that were undergoing changes in design and low-dimensional representations of multivariate samples may elucidate the extent of parallelism in related lineages. Thus the first role of biometry in explanatory investigations of shell design is to quantify the form of structures and to map their changes. This role is concerned with the collection of population data and in practice, if not intent, is closely allied to biometrical tasks in taxonomy.

The second role is deterministic rather than statistical. Understanding of complex relationships among structures, suggested by population studies, is assisted by study of models in which variables can be investigated under known conditions. Physical models have been constructed (Thompson, 1952) but geometrical models, able to be generated from a set of parameters and represented graphically in two or three dimensions, are probably more informative and more suitable for experimentation. There is no dilemma over choice of a first model. Thompson (1952, p. 854) wrote "it is obvious enough that the spiral shells of the Foraminifera closely resemble true logarithmic spirals" and referred to measurements by earlier workers that substantiated this statement. Hofker (1968) and Berger (1969) invoked the logarithmic spiral as a model to which many foraminiferal shells closely conform. Essential parameters required for modelling of logarithmic spiral shell are the shape of the chamber generating curve (the chamber profile is the realized segment of this curve), the rate of increment of chambers in size, the rate of translation of the chamber sequence along the coiling axis and the distance of chambers from the coiling axis. Construction of models is discussed, generally by Raup (1966), and specifically for foraminifera by Berger (1969).

Thompson (1952, p. 789) noted the simplicity of the model "the logarithmic spiral is but a plotting in polar coordinates of *increase by compound interest*". For example, a growth constant $a = d_{i+1}/d_i$ computed for linear measurements on successive expanding chambers is equivalent to $1 + A$ where A is the decimal interest rate per chamber. Suppose that chamber volume reflects protoplasmic volume. Growth by simple interest implies that increment in protoplasmic volume arises from a fixed volume, whereas compound growth suggests no differentiation into growth generating and growth generated portions. It may thus be appropriate as an initial model for growth of a protozoan. However, while it points to simple cytoplasmic organization, the logarithmic spiral model is one of unrestrained growth and maintenance of geometric similarity if values for parameters are unchanged during ontogeny. Under these conditions it implies, for example, that physiological processes are not bounded by surface:volume relationships and that skeletal structures can increment in size without impairing mechanical efficiency. In few organisms is such uncoupling of size effects possible. Gould (1971) remarked that geometrical similarity is a problem rather than an expectation. Metazoan growth curves may show an early exponential phase but this is usually followed by inflexion and a late ontogenetic phase in which rate of growth gradually declines (Bertalanffy, 1960). Moreover, size-required changes in shape of skeletal structures to maintain efficiency during growth are widespread in metazoans (Gould, 1966). While logarithmic spiral models of foraminifera (Fig. 36) in which parameters are fixed throughout ontogeny (Berger, 1969) may be informative, there will be a need to experiment with models in which parameters change value during growth. Predictably these will be required to study the advantages that accrue from ontogenetic changes in shell architecture so prominent in orbitoidal taxa and in smaller taxa that show marked, often abrupt, reduction in number of chambers per revolution. In orbuline shells the turbinate spiral formed by earlier chambers is enveloped almost completely by the last chamber. I have experimented with some logarithmical spiral models of these shells. Fixed parameter models can achieve an orbuline condition only if the relation of chamber centres to coiling axis is such that each successive chamber completely envelops its predecessors. The model in Fig. 37 (centre) is more similar to an orbuline shell than that on the left because chamber centres are closer to the coiling axis. But if rate of increment in chamber dimension is permitted to increase at the final growth stage, an orbuline shell is readily modelled (Fig. 37, right).

Orbuline or quasi-orbuline shells of the type modelled in Fig. 37 (right) arose in at least two Tertiary planktonic lineages. Thus an explanatory account of the adaptation might, for example, proceed to evaluate the hydromechanical efficiency of various orbuline models. Rate of passive sinking could be calculated. The most efficient models (lowest rates)

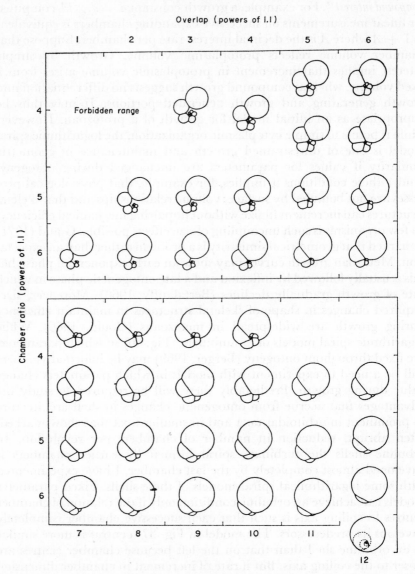

Fig. 36. Planispiral models of planktonic foraminifera with circular chamber profiles drawn for various values of chamber expansion and overlap. Forbidden range refers to combinations of parameters that produce chambers lacking contact with their predecessors and therefore not accretionary. Parameters for each model are constant throughout ontogeny. (After Berger, 1969).

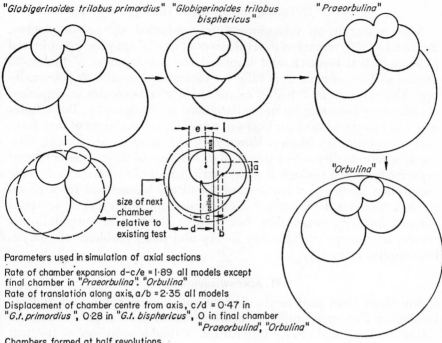

"Globigerinoides trilobus primordius" "Globigerinoides trilobus bisphericus" "Praeorbulina"

"Orbulina"

size of next chamber relative to existing test

Parameters used in simulation of axial sections

Rate of chamber expansion d−c/e = 1·89 all models except final chamber in "Praeorbulina". "Orbulina"

Rate of translation along axis, a/b = 2·35 all models

Displacement of chamber centre from axis, c/d = 0·47 in "G.t. primordius", 0·28 in "G.t. bisphericus", O in final chamber "Praeorbulina", "Orbulina"

Chambers formed at half revolutions

Generating curve circular

Fig. 37. Models of shells in the *Orbulina* lineage. Those at left and centre have parameters fixed throughout ontogeny; for those on right, rate of chamber expansion and displacement of chamber from axis are changed at the final growth stage. (Unpublished work by author).

may then be compared with the organization of actual shells. Close similarity would support the hydromechanical hypothesis as a possible explanation of the adaptation.

This section has been phrased largely in terms of what might be accomplished in analysis of shell design rather than what has been done. The merits of explanatory studies of shell design, and the role of biometry, cannot yet be demonstrated from the foraminiferal literature. Progress in this field seems to demand a change in outlook of foraminiferal workers, a willingness to ask "why" questions and to extend research beyond description. Can we really admit as a principle, as did Smout (1954), that the final shape of the shell is of little biological importance without asking "why"? Biometry should serve as a catalyst in stimulating explanatory studies because changes in shell architecture are made more explicit by measurement and relative growth analyses than by qualitative inspection. The curious will ask "why this *change* of shape?" (Gould, 1968).

D. Testing the Synthetic Theory

Most research on foraminifera is conducted by paleontologists. Rudwick (1968) remarked that the needs of stratigraphy have dominated paleontological research and impeded the development of biological attitudes. This assessment is valid for biometrical studies. A potentially powerful role can be filled by quantitative methodologies in empirical classificatory research, an immediate need in stratigraphy. But a fuller role will emerge as students begin to probe for explanations of shell form and of evolutionary history. When such explanations are sought, then the synthetic theory offers theses that, because of their coherence and veracity for much biological data, merit priority in testing. I advocate population studies as a cornerstone for future biometrical research on the foraminiferal shell, not as a metaphysical foundation to be accepted without further examination, but because the variable population is central to modern evolutionary theory and poses problems of analysis that require biometry.

VI. Acknowledgments

Assistance from staff in the library and photographic unit at N.Z. Geological Survey is gratefully acknowledged. Several paleontological colleagues suggested relevant literature. Valuable criticism of the text was provided by the editors and referee.

VII. References

Adams, C. G. (1957). *Micropaleontology* **3**, 205–226.

Alexander-Marrack, P. D., Friend, P. F. and Yeats, A. K. (1971). *In* "Data Processing in Biology and Geology" (J. L. Cutbill, ed.), pp. 1–16. Academic Press, London.

Anderson, T. W. (1958). "An Introduction to Multivariate Statistical Analysis". Wiley, New York.

Angell, R. W. (1967a). *J. Protozool.* **14**, 299–307.

Angell, R. W. (1967b). *J. Protozool.* **14**, 566–574.

Arnaud-Vanneau, A. (1968). *Geol. Alpine* **44**, 27–48.

Arni, P. (1967). *Micropaleontology* **13**, 41–54.

Bandy, O. L. (1960). *J. Paleont.* **34**, 671–681.

Bandy, O. L. (1968). *G. Geol.* (Bologna) ser. 2, **35**, 277–290.

Banner, F. T. and Blow, W. H. (1965). *Contr. Cushman Fdn foramin. Res.* **16**, 105–115.

Bartlett, M. S. (1949). *Biometrics* **5**, 207–212.

Bé, A. W. H. (1968). *Science* **161**, 881–884.

Beckner, M. (1959). "The Biological Way of Thought". Columbia University Press, New York.

Berger, W. H. (1969). *J. Paleont.* **43**, 1369–1383.

Berger, W. H. (1971). *Mar. Geol.* **11**, 325–358.

Berggren, W. A. (1960). *Stockh. Contr. Geol.* **5**, 41–108.

Berggren, W. A. (1971). *In* "The Micropalaeontology of Oceans" (B. M. Funnell and W. R. Riedel, eds.), pp. 693–809. Cambridge University Press, Cambridge.

Berggren, W. A. and Kurtén, B. (1961). *Stockh. Contr. Geol.* **8**, 1–16.

Berggren, W. A., Olsson, R. K. and Reyment, R. A. (1967). *Micropaleontology* **13**, 265–288.

Bertalanffy, L. von (1960). *In* "Fundamental Aspects of Normal and Malignant Growth" (W. W. Nowinski, ed.), pp. 137–259. Elsevier, Amsterdam.

Bettenstaedt, F. (1958). *Paläont. Z.* **32**, 115–140.

Bettenstaedt, F. (1962). *Mitt. geol. St. Inst. Hamb.* **31**, 385–460.

Bizon, G. and Bizon, J. J. (1971). *In* "Proceedings of the Second Planktonic Conference, Roma 1970" (A. Farinacci, ed.) Vol. 1, pp. 85–98. Edizioni Tecnoscienza, Rome.

Blackith, R. E. and Reyment, R. A. (1971). "Multivariate Morphometrics". Academic Press, London.

Blackmon, P. D. and Todd, R. (1959). *J. Paleont.* **33**, 1–15.

Blondeau, P. (1965). *Bull. Soc. géol. Fr.* **7**(vii), 268–272.

Blow, W. H. (1969). *In* "Proceedings of the First International Conference on Planktonic Microfossils, Geneva 1967" (P. Brönnimann and H. H. Renz, eds.) Vol. 1, pp. 199–422. Brill, Leiden.

Blum, H. (1973). *J. theor. Biol.* **38**, 205–287.

Bock, W. J. (1960). *Evolution* **14**, 130–132.

Bolli, H. M. (1950). *Contr. Cushman Fdn foramin. Res.* **1**, 82–89.

Bolli, H. M. (1951). *Contr. Cushman Fdn foramin. Res.* **2**, 139–143.

Bolli, H. M. (1971). *In* "The Micropalaeontology of Oceans" (B. M. Funnell and W. R. Riedel, eds.), pp. 639–648. Cambridge University Press, Cambridge.

Bradshaw, J. S. (1957). *J. Paleont.* **31**, 1138–1147.

Bretsky, S. S. (1970). *Bull. geol. Instn Univ. Upsala* n.s. **2**, 5–23.

Bretsky, S. S. (1971). *Syst. Zool.* **20**, 204–222.

Brönnimann, P., and Resig, J. (1971). *In* "Initial Reports of the Deep Sea Drilling Project" Vol. 7, pp. 1235–1469. U.S. Govt Printing Office, Washington.

Buchanan, J. B. (1960). *J. Linn. Soc.* **44**, 270–277.

Burma, B. H. (1942). *J. Paleont.* **16**, 739–755.

Burma, B. H. (1948). *J. Paleont.* **22**, 725–761.

Burnaby, T. P. (1966). *Biometrics* **22**, 96–110.

Buzas, M. A. (1966). *J. Paleont.* **40**, 585–594.

Camin, J. H. and Sokal, R. R. (1965). *Evolution* **19**, 311–326.

Caron, M. (1967). *Eclog. geol. Helv.* **60**, 47–79.

Carter, D. J. (1953). *J. Paleont.* **27**, 238–250.

Cassie, R. M. (1954). *Aust. J. mar. Freshwat. Res.* **5**, 513–522.

Cavalli-Sforza, L. L. and Edwards, A. W. F. (1967). *Evolution* **21**, 550–570.

Cifelli, R. (1969). *Syst. Zool.* **18**, 154–168.

Cifelli, R. (1971). *J. Foraminer. Res.* **1**, 170–177.

Cifelli, R. and Smith, R. K. (1970). *Smithson. Contr. Paleobiol.* **4**, 1–52.

Cock, A. G. (1963). *Genet. Res.* **4**, 167–192.

Cock, A. G. (1966). *Q. Rev. Biol.* **41**, 131–190.

Cole, W. S. (1960). *Micropaleontology* **6**, 133–144.

Cole, W. S. (1963). *Bull. Am. Paleont.* **46**, 157–185.

Cole, W. S. (1964). *Contr. Cushman Fdn foramin. Res.* **15**, 138–150.

Cole, W. S. (1966). *Bull. Am. Paleont.* **50**, 229–265.

Comfort, A. (1972). *Meanjin Q.* **31**, 351–355.

Cosijn, A. J. (1938). "Statistical Studies on the Phylogeny of some Foraminifera". Ijdo, Leiden.

Cosijn, A. J. (1942). *Leid. geol. Meded.* **73**, 140–171.

Cutbill, J. L. and Forbes, C. L. (1967). *Palaeontology* **10**, 322–337.

Day, N. E. (1969). *Biometrika* **56**, 463–474.

Dempster, A. P. (1969). "Elements of Continuous Multivariate Analysis". Addison-Wesley, Reading.

Douglass, R. C. (1970). *Prof. Pap. U.S. geol. Surv.* **643**, G1–G11.

Douglass, R. C. and Cotner, N. J. (1972). *J. Paleont.* **46**, 406–409.

Drooger, C. W. (1952). "Study of American Miogypsinidae". Vonk, Zeist.

Drooger, C. W. (1955). *Proc. K. ned. Akad. Wet.* ser. **B 58**, 415–433.

Drooger, C. W. (1956). *Proc. K. ned. Akad. Wet.* ser **B 59**, 458–469.

Drooger, C. W. (1963). *In* "Evolutionary Trends in Foraminifera" (G. H. R. von Koenigswald *et al.* eds.), pp. 315–349. Elsevier, Amsterdam.

Drooger, C. W. and Freudenthal, T. (1964). *Eclog. geol. Helv.* **57**, 509–528.

Drooger, C. W., Kaasschieter, J. P. H. and Key, A. J. (1955). *Verh. K. ned. Akad. Wet.* ser. **1 21**, 5–136.

Drooger, C. W., Marks, P. and Papp, A. (1971). *Utrecht micropaleont. Bull.* **5**, 1–137.

Dunbar, C. O. and Skinner, J. W. (1937). *Bull. Univ. Tex.* **3701**, 523–822.

Ehrlich, R. and Weinberg, B. (1970). *J. sedim. Petrol.* **40**, 205–212.

Eicher, D. L. (1960). *Bull. Peabody Mus. nat. Hist.* **15**, 1–126.

Ellerman, C. (1960). *Geol. Jb.* **77**, 645–710.

Ellis, C. H., Lohman, W. H. and Wray, J. L. (1972). *Colo. Sch. Mines Q.* **67**, 1–103.

Emiliani, C. (1969). *Micropaleontology* **15**, 265–300.

Ericson, D. B., Ewing, M. and Wollin, G. (1963). *Science* **139**, 727–737.

Feldmann, R. J. and Bryan, S. D. (1970). *In* "Picture Processing and Psychopictorics" (B. S. Lipkin and A. Rosenfeld, eds.), pp. 409–426. Academic Press, New York.

Freeman, H. (1970). *In* "Picture Processing and Psychopictorics" (B. S. Lipkin and A. Rosenfeld, eds.), pp. 241–266. Academic Press, New York.

Frerichs, W. E., Heiman, M. E., Borgman, L. E. and Bé, A. W. H. (1972). *J. foraminer. Res.* **2**, 6–13.

Freudenthal, T. (1969). *Utrecht micropaleont. Bull.* **1**, 1–208.

Freudenthal, T. (1972). *In* "IV Colloque Africain de Micropaléontologie" (Comite de Publication, eds.), pp. 144–162. Imp. Rectorate de Nice.

Friedman, H. P. and Rubin, J. (1967). *J. Amer. statist. Ass.* **62**, 1159–1178.

Ghose, B. K. (1966). *Contr. Cushman Fdn foramin. Res.* **17**, 104–108.

Ghose, B. K. (1970). *J. Geol.* **78**, 545–557.

Glaessner, M. F. (1955). *Micropaleontology* **1**, 3–8.

Gorbatschik, T. N. (1964). *Paleont. Zh.* **4(1964)**, 32–37.

Gould, S. J. (1966). *Biol. Rev.* **41**, 587–640.
Gould, S. J. (1967). *Evolution* **21**, 385–401.
Gould, S. J. (1968). *J. Paleont.* **42** (no. 5—suppl.), 81–98.
Gould, S. J. (1970). *Earth Sci. Rev.* **6**, 77–119.
Gould, S. J. (1971). *Am. Nat.* **105**, 113–136.
Gould, S. J. and Garwood, R. A. (1969). *Evolution* **23**, 276–300.
Gower, J. C. (1966). *Biometrika* **53**, 325–338.
Gower, J. C. (1972). *In* "The assessment of population affinities in man" (J. S. Weiner and J. Huizinga, eds.), pp. 1–24. Clarendon Press, Oxford.
Gower, J. C. and Ross, G. J. S. (1969). *Appl. Statist.* **18**, 54–64.
Grabert, B. (1959). *Abh. senckenb. naturforsch. Ges.* **498**, 1–71.
Gradstein, F. M. (1971). *In* "Proceedings of the Second Planktonic Conference, Roma 1970" (A. Farinacci, ed.) Vol. 1, pp. 583–590. Edizioni Tecnoscienza, Rome.
Griffiths, J. C. (1967). "Scientific Methods in Analysis of Sediments". McGraw Hill, New York.
Hansen, H. J. (1968). *Meddr dansk geol. Foren.* **18**, 345–348.
Hansen, H. J. (1970). *Meddr dansk geol. Foren.* **19**, 341–360.
Harris, D. (1968). *J. Anim. Ecol.* **37**, 315–319.
Hawkins, J. K. (1970). *In* "Advances in Information Systems Science" (J. T. Tou, ed.) Vol. 3, pp. 113–214. Plenum Press, New York.
Hay, W. W. (1971). *In* "Scanning Electron Microscopy, Systematic and Evolutionary Applications" (V. H. Heywood, ed.), pp. 123–143. Academic Press, London.
Haynes, J. (1962). *Contr. Cushman Fdn foramin. Res.* **13**, 90–97.
Hedley, R. H. (1964). *Int. Rev. Gen. & Exp. Zool.* **1**, 1–45.
Hodson, F. R., Kendall, D. G. and Tăutu, P. (1971). "Mathematics in the Archaeological and Historical Sciences". University Press, Edinburgh.
Hofker, J. (1951). "The Foraminifera of the Siboga Expedition part 3". Brill, Leiden.
Hofker, J. (1957). *Beih. geol. Jb.* **27**, 1–464.
Hofker, J. (1959). *Neues Jb. Geol. Paläont., Abh.* **108**, 239–259.
Hofker, J. (1968). *Publtiës natuurh. Genoot. Limburg* **18**, 5–135.
Hofker, J. (1972). *Rev. Esp. Micropaleontol.* **4**, 11–17.
Hofmann, G. W. (1967). *Geologica bav.* **57**, 121–204.
Hofmann, G. W. (1971). *N.Z. Jl Geol. Geophys.* **14**, 299–322.
Hottinger, L. (1960). *Schweiz. Palaeont. Abh.* **75/76**, 1–243.
Huxley, J. S. (1940). *In* "The New Systematics" (J. S. Huxley, ed.), pp. 1–46. Oxford, London.
Imbrie, J. (1956). *Bull. Am. Mus. nat. Hist.* **108**, 211–252.
Jardine, N. (1971). *Phil. Trans. R. Soc. Lond. Ser.* B **263**, 1–33.
Jardine, N. and Sibson, R. (1971). "Mathematical Taxonomy". Wiley, New York.
Jenkins, D. G. (1967a). *N.Z. Jl Geol. Geophys.* **10**, 1064–1078.
Jenkins, D. G. (1967b). *Micropaleontology* **13**, 195–203.
Jerković, L. (1969). *Godišnjak biol. Inst. Saraj.* **22**, 21–127.
Jolicoeur, P. (1963). *Growth* **27**, 1–27.
Jolicoeur, P. and Mosimann, J. E. (1960). *Growth* **24**, 339–354.

Kaesler, R. L. (1967). *In* "Essays in Paleontology and Stratigraphy, Raymond C. Moore Commemorative Volume" (C. Teichert and E. L. Yochelson, eds.), pp. 63–81. University of Kansas Press, Lawrence.

Kaesler, R. L. (1970). *In* "Proceedings of the North American Paleontological Convention" (E. L. Yochelson, ed.) Vol. 1, pp. 84–100. Allen Press, Lawrence.

Kaesler, R. L. and Waters, J. A. (1972). *Bull. geol. Soc. Am.* **83,** 1169–1178.

Kaever, M. (1960). *Geol. Jb.* **77,** 821–831.

Kendall, M. G. and Stuart, A. (1966). "The Advanced Theory of Statistics". Griffin, London.

Kennett, J. P. (1963). *N.Z. Jl Geol. Geophys.* **6,** 257–260.

Kennett, J. P. (1966). *Micropaleontology* **12,** 235–245.

Kennett, J. P. (1968a). *Science* **159,** 1461–1463.

Kennett, J. P. (1968b). *Micropaleontology* **14,** 305–318.

Khan, M. A. and Drooger, C. W. (1971). *Proc. K. ned. Akad. Wet. ser.* **B 74,** 97–121.

King, K. and Hare, P. E. (1972). *Science* **175,** 1461–1463.

Klaus, J. (1960). *Eclog. geol. Helv.* **53,** 285–308.

Klein, J. C. and Serra, J. (1972). *J. Microsc.* **95,** 349–356.

Koepnick, R. B. and Kaesler, R. L. (1971). *J. Paleont.* **45,** 881–887.

Kroboth, K. (1966). "Untersuchungen an *Citharina* d'Orb. (Foram.) aus dem Neokom Nordwest-Deutschlands". Präzis, Tubingen.

Kruskal, J. B. (1964a). *Psychometrika* **29,** 1–27.

Kruskal, J. B. (1964b). *Psychometrika* **29,** 115–129.

Ledley, R. S., Lubs, H. A. and Ruddle, F. H. (1972). *Comput. Biol. Med.* **2,** 107–128.

Lewis, K. B. and Jenkins, C. (1969). *Micropaleontology* **15,** 1–12.

Lewontin, R. C. (1966). *Syst. Zool.* **15,** 141–142.

Lindenberg, H. G. (1966). *Boll. Soc. paleont. ital.* **4,** 64–160.

Lindenberg, H. G. (1969). *In* "Proceedings of the First International Conference on Planktonic Microfossils, Geneva, 1967" (P. Brönnimann and H. H. Renz, eds.) Vol. 2, pp. 343–365. Brill, Leiden.

Loeblich, A. R. Jr. and Tappan, H. (1964). *In* "Treatise on Invertebrate Paleontology" (R. C. Moore, ed.) Part C, Protista 2. University of Kansas Press, Lawrence.

Lutze, G. F. (1964). *Contr. Cushman Fdn foramin. Res.* **15,** 105–116.

Lynts, G. W. and Pfister, R. M. (1967). *J. Protozool.* **14,** 387–399.

MacGillavry, H. J. (1956). *Verh. K. ned. geol.-mijnb. Genoot.* **16,** 296–308.

MacGillavry, H. J. (1959). *In* "El Sistema Cretacico" (L. B. Kellum and L. Benavides, eds.) Vol. 1, pp. 77–83. International Geological Congress, 20th, Mexico City.

MacGillavry, H. J. (1962). *Proc. K. ned. Akad. Wet. ser.* **B 65,** 429–458.

MacGillavry, H. J. (1965). *Proc. K. ned. Akad. Wet. ser.* **B 68,** 335–355.

MacGillavry, H. J. (1971). *Proc. K. ned. Akad. Wet. ser.* **B 74,** 206–238.

Malmgren, B. A. (1972). *Stockh. Contr. Geol.* **14,** 33–49.

Malmgren, B. A. and Kennett, J. P. (1972). *Micropaleontology* **18,** 241–248.

Mangin, J. P. (1966). *Eclog. geol. Helv.* **59,** 347–353.

Marcus, L. F. (1969). *Evolution* **23,** 301–307.

Marie, P. (1941). *Mém. Mus. natn. Hist. nat.*, Paris n.s. **12**, 95–258.

Marriott, F. H. C. (1971). *Biometrics* **27**, 501–514.

Maync, W. (1959). *Revue Micropaléont.* **2**, 153–172.

Maync, W. (1960). *Revue Micropaléont.* **3**, 103–118.

McGowran, B. (1971). *In* "Proceedings of the Second Planktonic Conference, Roma 1970" (A. Farinacci, ed.) Vol. 2, pp. 813–820. Edizioni Tecno-scienza, Rome.

Meulenkamp, J. E. (1969). *Utrecht micropaleont. Bull.* **2**, 1–168.

Michael, E. (1966). *Senckenburg. leth.* **47**, 411–459.

Mosimann, J. E. (1970). *J. Am. statist. Ass.* **65**, 930–945.

Mosteller, F. and Tukey, J. W. (1949). *J. Am. statist. Ass.* **44**, 174–212.

Mundry, E. (1972). *J. Int. Ass. Math. Geol.* **4**, 55–60.

Nagappa, Y. (1957). *Micropaleontology* **3**, 393–398.

Narasimhan, R. (1970). *In* "Picture Language Machines: Proceedings of a Conference held at the Australian National University" (S. Kaneff, ed.), pp. 1–30. Academic Press, London.

Neurath, P. W., Bablouzian, B. L., Warms, T. H. and Serbagi, R. C. (1966). *Ann. N.Y. Acad. Sci.* **128**, 1013–1028.

Newell, N. D. (1947). *Evolution* **1**, 163–171.

Nicol, D. (1944). *J. Paleont.* **18**, 172–185.

Noël, D. (1965). "Sur les Coccolithes du Jurassique Européen et d'Afrique du Nord". Éditions du Centre National de la Research Scientifique, Paris.

O'Herne, L. and Vlerk, I. M. van der (1971). *Boll. Soc. paleont. ital.* **10**, 3–18.

Olson, E. C. and Miller, R. L. (1951). *Evolution* **5**, 325–338.

Olson, E. C. and Miller, R. L. (1958). "Morphological Integration". University of Chicago Press, Chicago.

Olsson, R. K. (1971). *Trans. Gulf-Cst Ass. geol. Socs* **21**, 419–432.

Olsson, R. K. (1972). *Eclog. geol. Helv.* **65**, 165–184.

Olsson, R. K. (1973). *J. Paleont.* **47**, 327–329.

Orr, W. N. (1967). *Science* **157**, 1554–1555.

Orr, W. N. (1969). *Micropaleontology* **15**, 373–379.

Ozawa, T. (1970). *Geol. Palaeontol. Southeast Asia* **8**, 19–41.

Pessagno, E. A. (1964). *Micropaleontology* **10**, 217–230.

Pessagno, E. A. (1967). *Palaeontogr. am.* **5**, 245–445.

Piper, D. J. W. (1971). *In* "Data Processing in Biology and Geology" (J. L. Cutbill, ed.), pp. 97–103. Academic Press, London.

Pitcher, M. (1966). *In* "Computer Applications in the Earth Sciences: Colloquium on Classification Procedures" (D. F. Merriam, ed.), pp. 30–41. State Geological Survey of Kansas, Lawrence.

Pratt, V. (1972a). *Br. J. Phil. Sci.* **23**, 305–327.

Pratt, V. (1972b). *J. theor. Biol.* **36**, 581–592.

Prewitt, J. M. S. and Mendelsohn, M. L. (1966). *Ann. N.Y. Acad. Sci.* **128**, 1035–1053.

Quilty, P. G. (1969). *J. Proc. R. Soc. West. Aust.* **52**, 41–58.

Rao, C. R. (1952). "Advanced Statistical Methods in Biometric Research". Wiley, New York.

Raup, D. M. (1966). *J. Paleont.* **40**, 1178–1190.

Rensch, B. (1959). "Evolution above the species level". Methuen, London.

Reyment, R. A. (1959). *Stockh. Contr. Geol.* **3**, 1–57.
Reyment, R. A. (1960). *Stockh. Contr. Geol.* **7**, 1–238.
Reyment, R. A. (1961). *Stockh. Contr. Geol.* **8**, 17–26.
Reyment, R. A. (1962). *Biometrics* **18**, 1–11.
Reyment, R. A. (1966). *Eclog. geol. Helv.* **59**, 319–337.
Reyment, R. A. (1968). *Publ. palaeont. Instn Univ. Upsala* **74**, 1–11.
Reyment, R. A. (1969). *Bull. geol. Instn Univ. Upsala* n.s. **1**, 75–81.
Roberts, T. G. (1953). *Mem. geol. Soc. Am.* **58**, 167–226.
Rohlf, F. J. (1972). *Syst. Zool.* **21**, 271–280.
Rozlozsnik, P. (1929). *Geologica hung.* (Seria palaeontologica) **2**, 1–248.
Rudwick, M. J. S. (1964). *Br. J. Phil. Sci.* **15**, 27–40.
Rudwick, M. J. S. (1968). *J. Paleont.* **42** (no. 5—suppl.), 35–49.
Sanderson, G. A. (1971). *J. sedim. Petrol.* **41**, 316–320.
Schaub, H. (1951). *Schweiz. palaeont. Abh.* **68**, 1–222.
Schaub, H. (1963). *In* "Evolutionary Trends in Foraminifera" (G. H. R. von Koenigswald *et al.* eds.), pp. 282–297. Elsevier, Amsterdam.
Scott, G. H. (1965). *Palaeont. Bull., Wellington* **38**, 1–48.
Scott, G. H. (1966a). *N.Z. Jl Geol. Geophys.* **9**, 203–211.
Scott, G. H. (1966b). *N.Z. Jl Geol. Geophys.* **9**, 513–540.
Scott, G. H. (1968). *N.Z. Jl Geol. Geophys.* **11**, 356–375.
Scott, G. H. (1970). *Micropaleontology* **16**, 385–398.
Scott, G. H. (1971). *Rev. Esp. Micropaleontol.* **3**, 283–292.
Scott, G. H. (1972a). *Micropaleontology* **18**, 81–93.
Scott, G. H. (1972b). *N.Z. Jl Geol. Geophys.* **15**, 287–295.
Seal, H. L. (1966). "Multivariate Statistical Analysis for Biologists". Methuen, London.
Simpson, G. G. (1953). "The Major Features of Evolution". Columbia University Press, New York.
Simpson, G. G. (1961). "Principles of Animal Taxonomy". Oxford, London.
Simpson, G. G., Roe, A. and Lewontin, R. C. (1960). "Quantitative Zoology". Harcourt Brace, New York.
Sliter, W. V. (1970). *Contr. Cushman Fdn foramin. Res.* **21**, 87–99.
Smith, P. B. (1963). *Prof. Pap. U.S. geol. Surv.* **429**, A11–A39.
Smout, A. H. (1954). "Lower Tertiary Foraminifera of the Qatar Peninsula". British Museum (Natural History), London.
Sneath, P. H. A. (1967). *J. Zool.* **151**, 65–122.
Snee, R. D. and Andrews, H. P. (1971). *Appl. Statist.* **20**, 250–258.
Sokal, R. R. and Rohlf, F. J. (1969). "Biometry". Freeman, San Francisco.
Sokal, R. R. and Sneath, P. H. A. (1963). "Principles of Numerical Taxonomy". Freeman, San Francisco.
Souaya, F. J. (1961). *Proc. K. ned. Akad. Wet. ser.* B **64**, 665–705.
Sprent, P. (1972). *Biometrics* **28**, 23–37.
Sylvester-Bradley, P. C. (1971). *In* "Scanning Electron Microscopy, Systematic and Evolutionary Applications" (V. H. Heywood, ed.), pp. 95–111. Academic Press, London.
Takayanagi, Y. Niitsuma, N. and Sakai, T. (1968). *Sci. Rep. Tôhoku Univ.* (Geol.) **40**, 141–170.
Tan, S. H. (1932). *Wet. Meded. Dienst Mijnb. Ned.-Oost-Indië* **19**, 3–194.

Tan, S. H. (1936). *Ing. Ned.-Indië* **3**, 45–61.
Tan, S. H. (1939). *Ing. Ned.-Indië* **6**, 93–97.
Theyer, F. (1971). *Antarct. Res. Ser.* **15**, 309–313.
Thompson, D. W. (1952). "On Growth and Form". Cambridge University Press, Cambridge.
Thompson, M. L. (1954). *Paleont. Contr. Univ. Kans.* (Protozoa) **5**, 1–226.
Towe, K. M. and Cifelli, R. (1967). *J. Paleont.* **41**, 742–762.
Ujiié, H. (1963). *Bull. natn. Sci. Mus., Tokyo* **6**, 378–404.
Ujiié, H. (1966). *Bull. natn. Sci. Mus., Tokyo* **9**, 413–430.
Ujiié, H. and Oshima, K. (1960). *Sci. Rep. Tokyo Kyoiku Daig.*, **7**, 105–116.
Van Valen, L. (1965a). *Evolution* **19**, 347–349.
Van Valen, L. (1965b). *Q. Rev. Biol.* **40**, 188–189.
Vašíček, Z. and Jičín, R. (1972). *Syst. Zool.* **21**, 91–96.
Vlerk, I. M. van der (1959). *Q. Jl geol. Soc. Lond.* **115**, 49–63.
Vlerk, I. M. van der (1963). *Micropaleontology* **9**, 425–426.
Vlerk, I. M. van der (1968). *Micropaleontology* **14**, 334–338.
Vlerk, I. M. van der and Bannick, D. D. (1969). *Proc. K. ned. Akad. Wet. ser.* **B 72**, 169–174.
Vlerk, I. M. van der and Gloor, H. (1968). *Genetica* **39**, 45–63.
Vlerk, I. M. van der and Postuma, J. A. (1967). *Proc. K. ned. Akad. Wet. ser.* **B 70**, 391–398.
Wade, M. (1964). *Micropaleontology* **10**, 273–290.
Walker, G. F. and Kowalski, C. J. (1972). *Comput. Biol. Med.* **2**, 235–249.
White, J. F. and Gould, S. J. (1965). *Am. Nat.* **99**, 5–18.
Wiles, W. W. (1967). *In* "Progress in Oceanography" (M. Sears, ed.) Vol. 4, pp. 153–160. Pergamon, Oxford.
Wolanska, H. (1959). *Acta palaeont. pol.* **4**, 27–59.
Wright, R. M. and Switzer, P. (1971). *J. Int. Ass. Math. Geol.* **3**, 297–311.
Zobel, B. (1968). *Geol. Jb.* **85**, 97–122.

Field and Laboratory Techniques for the Study of Living Foraminifera

ZACH M. ARNOLD

Department and Museum of Paleontology, University of California, Berkeley

I. Historical Introduction	. . . . . .	154
II. Choice of Species for Study	. . . . .	159
III. Field Techniques and Equipment	. . . . .	162
A. General	. . . . . .	162
1. Basic Methods of Collecting	. . . .	162
2. Choice of Collecting Site; Sources of Specimens	. .	163
3. Miniaturization of Equipment	. . . .	164
B. Collecting Tools	. . . . . .	166
1. Scrapers and Knives	. . . . .	166
2. Dredges	. . . . .	168
C. Field Examination	. . . . .	170
D. Preliminary Concentration (Field)	. . . .	173
E. Transporting to Laboratory	. . . .	176
F. Miscellany	. . . . .	177
IV. Pre-Culture Maintenance and Preparation	. . .	178
A. Interim Care	. . . . .	178
B. Examination and Evaluation of Field Collections	. .	178
1. General Condition of Specimens	. . .	178
2. Distinguishing Living From Dead Specimens	. .	180
C. Concentrating Specimens	. . . .	185
V. Culture Techniques and Equipment	. . . .	189
A. Types of Culture	. . . . .	189
B. Culture Containers	. . . . .	190
1. Aquaria and Large-Volume Containers	. .	190
2. Small-Volume Containers	. . . .	191
C. Culture Storage	. . . . .	193
D. Nutrient and Culture Media	. . . .	193
E. Food and Feeding	. . . . .	195
F. Other Culture Conditions	. . . .	196
1. Salinity	. . . . .	196
2. Hydrogen-ion Concentration	. . .	196
3. Oxygen	. . . . .	197
4. Light	. . . . .	197
5. Temperature	. . . . .	197
6. Associated Organisms	. . . .	199
7. Substratum	. . . .	200
G. Subculturing	. . . . .	201
H. Harvesting	. . . . .	203
I. Observation	. . . . .	203
VI. References	. . . . . .	204

I. Historical Introduction

The laboratory study of living foraminifera, long an exacting technical challenge for the protozoologist, has been vigorously stimulated during the past few decades by an increasing paleontological and geological need for the type of fundamental biological information that could facilitate the interpretation of the group's extensive fossil record. Widespread misconceptions about the difficulty of establishing these organisms in the laboratory, however, persisted until the resurgence of interest in the group after World War II. Although difficulties are, admittedly, numerous, and the vast majority of species have either never been brought into the laboratory in living condition or have proved discouragingly unco-operative once there, it has long been known that several common inshore or shallow-water species can be maintained with the simplest of facilities and crudest of procedures.

The approach employed by most students in cultivating foraminifera has been empirical and pragmatic, the study of culture methods generally not the end itself but merely the means to other biologically or paleontologically significant ones. That this should be true is not surprising when one considers several inherent properties that render the foraminifera as a group more difficult to handle experimentally than numerous other popular laboratory protozoa, principal among which are:

a. Their relatively slow rate of reproduction and commensurately long life span, often with the additional complication of an alternation of generations.

b. Their comparatively subtle vegetative activities and the consequent difficulty of accurately assessing their general condition.

c. The relatively slow rate at which bionomical knowledge essential to the efficient procurement of adequate stocks has been disseminated in biological circles.

d. The remoteness of many laboratories from the sea and easy access to the living organisms.

e. Long-standing misconceptions about the maintenance of marine cultures.

f. The difficulty of achieving fertilization in non-plastogamic species, and the numerous hazards facing their gametes under laboratory conditions.

g. The general lack of information about the ecology of laboratory cultures, particularly nutritional requirements and inter-organismal effects.

The study of cultivation procedures for foraminifera has not been particularly innovative, the methods and equipment largely mere adaptations from other areas of protozoology and marine biology, but the success achieved by these means during the past century attests to the efficacy of such pragmatism, however superficial and unquestioning the application might have seemed at the time. Viewing the Order Foraminiferida as a whole, however, the success has been very limited. This emphasizes the need for the more extensive experimental study of the natural and laboratory requirements of these organisms.

The purpose of the present account is to review and describe some of the simple procedures and equipment known to be effective in raising certain species of foraminifera. The reality of the continuing need for an expanded understanding of biological principles basic to the successful domestication of additional species from the broad spectrum of untested or recalcitrant types cannot be gainsaid, but when pressed to choose between that, on the one hand, which is ideal and theoretically desirable but still unachieved and that, on the other, which, though imperfect, is effective within the limits of its imperfections, one must rely on a sense of values that accurately appraises the levels of refinement and the limits of significance within which one is operating. Thus, axenic cultures, though obviously desirable from some viewpoints and admittedly essential to certain types of nutritional and other physiological investigations, are patently luxurious rather than essential in the solution of many equally fundamental biological problems of test morphology, ontogeny, life cycles, cytology, variability, and patterns of inheritance. When one turns to problems of potential interest to paleontologists, the levels of utilizable refinement and degrees of significance are even further relaxed, in consequence of the general blurring of the record of life and environments of past times by the processes of death, burial, post-depositional change, and fossilization. Unless a particular physiological, or functional, attribute can be shown to have a specific and unalterable morphological or biochemical attribute that can be recognized in the organism's fossil remains, its highly refined definition or characterization by the biologist has little meaningful value in paleontological interpretation. To date such attributes are notoriously few.

Nineteenth and early twentieth century interest in raising foraminifera in the laboratory seems principally to have been biological in its stimulus and essentially theoretical, or academic, in its approach, although concurrent studies by geologists of fossil and recent tests alike undoubtedly played an important catalytic role. It is easier to cross interdisciplinary lines in theory than in practice, because the latter requires technical mastery as well as mere theoretical knowledge. In the nineteenth century, geological interest in and knowledge of foraminiferal biology was more theoretical and academic than applied, and the differentiation between geology and biology was not as clear as it subsequently came to be, but with the discovery of the value of fossil foraminifera in correlating sedimentary strata and interpreting past environmental conditions (this information being of great value in the search for petroleum) the stage

was set for the generation of an even more powerful geological stimulation to biological inquiry. T. Wayland Vaughan, the geologist who early
encouraged J. A. Cushman in his successful application of foraminiferology to petroleum exploration (Kleinpell, 1971), later, as Director of
the Scripps Institution of Oceanography, employed E. H. Myers as a
student technician preparing sections of orbitoids, an association which
probably helped alert Myers, a biologist, to the need for closer coopera-
tion between biologists and paleontologists in the study of foraminifera.
Such interplay between biological and geological thought helped prepare
the way for the post-war upsurge in foraminiferal geobiology.

The principal emphasis of laboratory studies prior to the time of the
Second World War was on life cycles and cytology, this largely reflecting
either the predominance of biologists in the field or the response of
biologists to the geologists' expressed desire for more basic biological
information (for example, within the area of variation potentials, with
important taxonomic implications). As geologists began to recognize the
value of foraminifera in paleoecological interpretation and to appreciate
the necessity of carefully distinguishing faunal differences attributable
to environmental factors from those attributable to evolutionary processes, they stimulated interest in ecological studies but found so few
biologists willing or prepared to undertake them that they were often
forced to do so themselves.

The influx into a biological arena of workers whose background
training and interests are basically geological has undoubtedly produced
much biologically naïve or even unsound research, but it has also
injected into the field of foraminiferal biology fresh interest and imaginative challenge of a type not to be expected in geologically naïve biologists. At the same time it has made geologists much more aware of the
complexity of living systems and the difficulties inherent not only in the
experimental study of living organisms but in the analogical transfer of
knowledge so gained to the interpretation of life of past times. A principal
attribute of geologically manned research in foraminiferal biology has
been its empiricism, this particularly evident in the techniques of
collecting and culturing the organisms, but the cumulative wealth of
geological and biological experience has contributed much to the broad
and solid foundation of particular truths from which a cohesive body of
general truths—principles—can ultimately be induced. The time now
seems ripe for logarithmic growth in our knowledge of the biological
fundamentals of Foraminiferology.

In the absence of extensive historical research into the subject, one can
but speculate on the influence such aquariists as P. H. Gosse and W. A.
Lloyd might have had on the study of living foraminifera in Europe's
academic institutions. Gosse (1856, p. 278) reported that in his tanks
"the Foraminifera increase with me abundantly", but his writings on

marine aquaria post-date the earliest biological observations [Dujardin (1835) and Gervais (1847)] on living specimens from the laboratory. It seems unlikely, too, that Schultze's principal studies (1854, 1860) could have been influenced by them, although the first of the large public marine aquaria, that of the Zoological Society of London (Yonge, 1949, p. 24), was established in 1853 with collections made by Gosse himself. With the lively interest in home aquaria stimulated by the popular writings of Gosse (1856), the ready availability of equipment and specimens from such massive stocks as Lloyd's London ones (1858), and the successful establishment of public aquaria at the Jardin d'Acclimatation in Paris (1860), Hamburg (1864), Brussels (1868), Berlin (1869), the Crystal Palace in London (1871) [see Lloyd's account of this, 1875], and the Zoological Station in Naples (1872), the stage was certainly set for the adoption of advanced techniques in aquarium management by the academician.

Perhaps, though, the early students of living foraminifera considered complicated methods for maintaining these organisms unnecessary. In practice, this certainly proved to be so for several species, as Dujardin early realized. He found (1835, footnote on p. 344) that "Des Milioles, des Vorticiales et des Cristellaires" from Toulon lived in "un bocal contenant quatre onces d'eau de mer . . . non renouvelée" from July 2nd to August 20th, when he had to leave them. Later he found that specimens collected on November 6th from the Manche coast were still alive in his Paris laboratory at the end of February ("dans l'eau de mer renouvelée"). He considered their hardiness in the laboratory to be an indication of the simplicity of their organization, since they survived when more complex animals (isopods, annelids, and brachiopods) succumbed. He astutely observed that "à la vérité leur habitation dans les touffes de corallines les met naturellement déjà dans une eau moins pure et presque dans les conditions d'une veritable infusion."

Max Schultze, too, seemed to rely, with a modicum of success at least, on this hardiness. He (1856) carried Adriatic specimens from Ancona and Venice to Greifswald and kept *Polystomella strigilata*, some *Gromia*, *Rotalia*, and miliolids alive from September and October to July of the following year "in meinen Gläsern" in which the water was changed only once. Later (1860) he kept a small number of "*Gromiae* (*G. dujardinii*)", "*Miliolidae*", and "*Rotalidae*" alive in a bottle with sea-water and a little sand from the autumn of 1857 to June of 1859 when he was able to observe reproduction in living specimens of a form similar to *Rotalina nitida* Williamson.

F. E. Schulze (1875) found large numbers of *Spiroloculina hyalina* in "einem Aquarium, welcher seit etwa einem Jahre mit Pflanzen und Thieren von den Pfählen der Warnowmündung besetzt war . . .", but, unfortunately, by the time he began to examine them closely no living specimens could be found. Now known to be a hardy species for laboratory use (Arnold, 1964, Lee *et al.*, 1966, and Muller and Lee, 1969), the creature had undoubtedly multiplied extensively between the time that Schulze collected the algae and eventually examined the microfauna. He did not specify the type of aquarium he employed, but, presumably, it was at least larger than the "Gläser" of earlier workers.

By the time Lister and Schaudinn began publishing their fundamental observations on foraminiferal biology, both the Naples and the Plymouth laboratories were established (1872 and 1888 respectively) and had begun to exert an important influence on the course of marine biology generally, so it is reasonable to suppose that foraminiferal biologists, by the turn of the century, at least, had begun to think in terms of circulating aquaria for their laboratory populations. In 1894(a) Schaudinn was able to report on observations made upon *Gromia oviformis* (called *Hyalopus* n.g. by him, and now known not to be a true foraminifer) living in the marine aquarium at the Berlin Zoological Institute. Schulze, having moved from Bonn to Berlin, was now Schaudinn's professor, so it might well be that the distinction of first employing well-established institutional marine aquaria in the sustained scientific study of foraminifera should be accorded either Schulze, who, however, did not publish any of his post-Bonn (*i.e.*, Berlin) studies, or Schaudinn, who published extensively until the time of his premature death (1906).

In spite of the growing awareness of marine biologists for the potential of aquaria in maintaining laboratory populations, students of foraminiferal biology generally have relied more on non-circulating and relatively small-volume containers for their more critical studies (e.g., Myers, 1937), although they have not seemed averse to availing themselves of specimens from aquaria when the opportunity arose. The reasons for this preference will shortly be detailed.

The challenge of applying the fruits of modern technological progress to the study of the foraminifera is an exciting one. The potentials of scanning and transmission electron microscopy, computer science, submersible vehicles, and underwater breathing apparatus for divers are being actively probed on a reasonably wide front, but surprisingly, other areas of rich promise and obvious relevance have been only superficially tapped. One of these is plastics technology, capable of opening many new horizons in the design of foraminifer-handling equipment in field and laboratory. Thirty years ago glass finger bowls, Boveri dishes, and Petri dishes, long used by protozoologists, were still the most efficient containers for most non-circulating culture systems, but today a diversified assortment of boxes and dishes injection-molded in clear plastics is readily available, and the imaginative search through supply catalogues often discloses containers easily adaptable to particular requirements and experimental endeavors. Biologists generally, however, seem unaware of the opportunities that exist for designing and producing containers and experimental gear from plastics through the use of simple, inexpensive equipment easily built in a workshop moderately well equipped with standard metal-working machine tools. With vacuum-forming equipment fabricated from materials costing but a few dollars and operated by a second-hand vacuum cleaner, the author has made a varied assortment of containers and tools that have proved most useful in field and laboratory alike, some of which are described in the following account. The need for well-designed equipment for growing and studying foraminifera efficiently and rapidly becomes apparent when one uses conventional protozoological equipment in working with these marine sarcodinians. Modern plastics technology, if imaginatively applied to the problems, can greatly simplify the chore and increase the biologist's productivity.

II. Choice of Species for Study

The species a paleontologist would select for biological study are not necessarily those a biologist would, because the two specialists have different interests and needs. The former would naturally prefer to learn more about the biology of species important to the interpretation of the fossil record. This generally means species having preservable hard parts, i.e., calcareous or agglutinated tests. Ideally, the species should be the same as its fossil counterpart or, in the case of one now extinct, its closest living relative. Since most paleontological inquiry is directed toward species from the offshore waters of the continental slope (the region of the order's maximum diversity), biological study, to be of greatest paleontological value, should concentrate on the living counterparts in such offshore environments. Those species that have been found to be of greatest value as stratigraphic markers or paleoecological indicators should be studied experimentally, and, in tune with current trends, particular attention should be given planktonic species. Moreover, paleontological need generally requires attention to species whose individual representatives are sufficiently large to be easily manipulated and studied in the type of practical micropaleontological investigations routinely applied to the solution of geological problems (particularly stratigraphic ones).

While the paleontologist would select fairly large offshore species with moderately heavily mineralized tests, the biologist generally prefers small inshore or intertidal ones with non-mineralized or weakly mineralized tests. The smaller species are generally easier to maintain, their life cycle and generation time are generally shorter, satisfactory kill-fixing is more easily accomplished, more specimens can be concentrated in a given volume of paraffin for microtomy, and, in almost every respect, they are technically less demanding. Intertidal forms are not only easier

TABLE 1
Successful and Promising Species of Foraminifera
for Laboratory Cultivation[1]

Foraminifer	Food Source	Reference
*Allogromia laticollaris	naviculoid diatoms and blue-green algae	Arnold, 1955
*Allogromia laticollaris	Chlorella	Schwab, 1970
*Allogromia sp.	Dunaliella parva; monoxenic on a pseudomonad	Lee and Pierce, 1963
Allogromia spp.	Tetraselmis chui	Hedley et al., 1973
Astrorhiza limicola	various metazoans including copepods, nematodes etc.	Buchanan and Hedley, 1960
Boderia turneri	Dunaliella praemolecta	Hedley et al., 1968
*Bolivina doniezi	Nitzschia angularis	Sliter, 1970
*Calcituba polymorpha	naviculoid diatoms and blue-green algae	Arnold, 1967
*Cibicides lobatulus	Chlamydomonas	Nyholm, 1961
Cornuspira lajollaensis	diatoms	Gougé, 1971
Discobotellina biperforata	diatoms	Stephenson and Rees, 1965
*Discorbinopsis aguayoi	pennate diatoms and blue-green algae	Arnold, 1954b
Elphidium crispum	diatoms	Jepps, 1942
Elphidium crispum	Phaeodactylum	Murray, 1963
Glabratella sulcata	pennate diatoms	Grell, 1958
Globigerina bulloides	planktonic algae	Adshead, 1967
Haliphysema tumanowiczii	diatoms	Hedley, 1958
*Heterostegina depressa	—	Röttger, 1972
Hippocrepinella alba	not specified	Nyholm, 1955
Iridia diaphana	not specified	Marszalek, 1969
Iridia lucida	diatoms	Le Calvez, J., 1938
Marginopora vertebralis	not specified	Ross, 1972
*Metarotaliella parva	diatoms	Weber, 1965
*Myxotheca arenilega	algal zoospores	Føyn, 1936b
*Myxotheca arenilega	Chlorella	Grell, 1972
Nemogullmia longevariabilis	not specified	Nyholm, 1956
*Patellina corrugata	diatoms	Grell, 1959
Planorbulina mediterranensis	diatoms and dried plankton	Le Calvez, J., 1938
*Rosalina floridana	diatoms	Freudenthal et al., 1963
Rosalina floridana	mixed algae	Angell, 1967
Rosalina floridana	diatoms	Chinn, 1972
*Rosalina leei	Dunaliella praemolecta	Hedley and Wakefield, 1967

[1] Species indicated by an asterisk (*) have been successfully established and maintained for long periods. The other species show promise of eventual successful establishment in the laboratory.

TABLE 1 (contd.)

Foraminifer	Food Source	Reference
Rosalina (=Discorbina) vilardeboana	algal zoospores	Føyn, 1936a
*Rotaliella heterocaryotica	Dunaliella	Grell, 1954
Rotaliella roscoffensis	diatoms	Grell, 1957
*Saccamina alba	diatoms	Hedley, 1962
Shepheardella taeniformis	not specified	Hedley et al., 1967
Spirillina vivipara	diatoms	Myers, 1936
*Spiroloculina hyalina	diatoms and blue-green algae	Arnold, 1964
Streblus (=Rotalia) beccarii tepida	Nitzschia; Chlamydomonas	Bradshaw, 1955

to procure, they are easier to maintain in the laboratory than are forms from deeper water. Moreover, with his general interest in the fundamentals of the organism's cytology and life cycle, the biologist would prefer to use species with relatively large nuclei and relatively few, but large, chromosomes. The disparity between the theoretical desiderata set by the paleontologist and the practical guidelines Nature sets for the biologist can hardly better be indicated than by the observation that for not a single paleontologically significant species has a recognizable mitotic or meiotic figure yet been published.

The past two decades have seen increased interest develop in other aspects of foraminiferal morphology—particularly test ultrastructure—and function that, though of fundamental importance to the paleontologist, require a less rigorous program of culture control and a less comprehensive understanding of the vagaries of the organism's life activities than does the analysis of nuclear, chromosomal, and life cycles. Much information of interest to biologists and use to paleontologists can be acquired from the study of species casually collected and not successfully established in the laboratory, so at this level the horizons of species choice for the biologist are far wider than when the limitations of true laboratory domestication are imposed.

The use of foraminifera as indicators of pollution in the biologist's expanding involvement with environmental-impact studies can well stimulate culture programs, because field investigations should be bolstered with comprehensive laboratory studies of physiological stresses, tolerances and responses, and the morphological correlates of the more readily monitored and experimentally controllable environmental variations. To be of real service to the ecologist, the foraminiferal biologist needs to expand his culture horizons to include the numerous species critical to pollution studies that have not yet been successfully domesticated.

In the present discussion, our basic concern is with establishing species successfully in the laboratory, so that they grow to maturity, reproduce at a reasonably normal rate, and thus provide populations that constitute a permanent and dependable source for further study. The species that have been maintained for a few to many years in various laboratories of which the author has been able to procure records are indicated by an asterisk in Table 1. The others are species that seem to show promise of successful establishment on a long-term basis but have not yet been reported as truly established.

III. Field Techniques and Equipment

A. General

1. BASIC METHODS OF COLLECTING

Access to the foraminifera of shallow waters and inshore areas may be by small boat, through the use of underwater breathing devices, or, if essentially shore-based, by swimming, wading and casting. The major types of sampling devices employed aboard ocean-going vessels may be made smaller and lighter and otherwise modified for use aboard small boats, where they are employed with a degree of success commensurate with the natural differences in distributional pattern of offshore and inshore foraminifera, but the ready availability of underwater breathing devices and other equipment has opened vast areas to direct collection. Some of the tools and methods used in intertidal collecting, the simplest and still one of the most effective of the direct methods, have successfully been adapted by divers for subtidal and deeper use, but they have also developed specialized ones for this more rigorous approach.

Through the proper use of suitably designed equipment an experienced and astute collector can procure abundant supplies of a challenging assortment of foraminiferal species easily, quickly, far less expensively than by scuba diving, ship and boat, and more safely than by diving. Moreover, inasmuch as species from intertidal and shallower waters are generally best adapted for the rigours of the changeable nearshore environment, they are usually more promising subjects for artificial cultivation than their deeper-water counterparts. Then, too, wave and current action often displace subtidal species into intertidal areas where they may beneficially be trapped and held to await garnering by the experienced wading or hip-boot collector.

There are still, however, certain troublesome and intriguing inshore areas not directly or safely accessible by the means so far mentioned. Dangerous surf-swept rocky shores, treacherous marshy expanses, quick sands, the outer reaches of extensive mud flats, all present hazards best circumvented by the use of remote-controlled sampling devices. Such areas may most safely be sampled by casting lightweight grapples, dredges, or nets from shore by means of the fisherman's surf-casting equipment or stronger propulsive devices, such as an archer's bow or cross-bow, sling or rubber catapults, or even explosive-powered devices like those used for line-throwing and rescue work by sailors and firemen. The vast potential of these devices, or radio-controlled modifications of model planes and boats—to say nothing of kites and balloons—has hardly been explored by the foraminifer collector.

The methods and equipment described herein are principally those developed for shore-based operations of the wading variety, but modifications for use in other types of operation should immediately be apparent.

2. CHOICE OF COLLECTING SITE; SOURCE OF SPECIMENS

Although foraminifera are widely distributed in intertidal areas, their abundance varies greatly from place to place and from time to time, so the experienced collector must use all available clues in his field search for rich sources of the particular species he seeks. In unusually favorable environments one can collect sediment, algae and marine grasses almost indiscriminately and be assured of finding many foraminifera, but more often some care, based on past experience and an understanding of the forces that govern intertidal distribution, pays dividends.

Generally speaking, beaches of quartz sand along exposed stretches of coastline in areas of temperate to cool or cold water are unpromising sites for rich collecting, because the development of deposits of sand-sized particles usually indicates vigorous wave or current action, the usual consequence of this being the destruction of foraminiferal tests. At certain times of the year, however, strandline deposits rich in foraminifera develop even on such sandy beaches, and within these, living specimens (or, better, survivors) may be numerous; but such deposits are often extremely ephemeral and hazardous to predict, hence unreliable. In warmer climates and carbonate environments, abundant foraminifera may be found in intertidal deposits of sand-sized particles, and during quiet seasons foraminifera from the quieter and deeper subtidal and offshore waters may invade the inshore areas and proliferate among the calcareous sand grains. In silicate environments, too, specimens may at times move inshore from deeper water and even be found attached to sand grains where their chances of survival are good until inclement weather begins grinding the grains with its waves.

Living foraminifera, like their fossil remains, are more frequently encountered in association with fine-grained sediments, ranging from fine sand to silt, than in coarser ones. The development of deposits within this size range in shallower waters and intertidal areas reflects less violent water movement and usually indicates a more protected environment, but other factors, such as dilution with fresh-water from streams and surface runoff, for example, complicate the picture and must be taken into consideration by the collector. With the exception of planktonic species (seldom encountered in intertidal waters except around oceanic islands or along open coastal stretches against which offshore winds can concentrate them) foraminifera are most commonly found crawling upon or attached to solid surfaces: rocks, sediment, and the bodies of plants and animals.

Rock surfaces, whether they be formed of igneous or metamorphic material or of consolidated sediments, are often a good source of attached foraminifera, their lithologic characteristics generally less important as a limiting factor than the degree of exposure to wave and current action. If the rocks are in calm, protected waters, both vagile and attached species can develop in profusion on practically any type of surface, but if the area is exposed to vigorous wave or current action, the composition, weathering characteristics, and attitudinal relations of the rocks become critical. In order to provide protection for the foraminifera, rocks so exposed must be of such composition as to permit ready attachment by animals and plants which will, in turn, provide the necessary haven, or they must so weather as to provide nooks and crannies wherein the protozoan can safely nestle. An easily degraded shale, for example, not only provides little stability for the attachment of either macro- or micro-organisms, but its rapid deterioration produces a turbidity that can be inimical to life. By contrast, the loss of each grain of a mildly indurated standstone or a granite from the parent rock to the crashing surf creates a new pit in which a foraminifer can find refuge. A good, but not infallible, clue to a source of foraminifera on exposed rock surfaces is the presence of a rich growth of algae (particularly of such closely-knit types as the corallines), hydroids, or masses of such closely-knit colonial invertebrates as the tufty, arborescent bryozoans. The most productive plants or animals are those that appear "dirty" to the naked eye, *i.e.*, bearing accumulations of particles within the fine-sand-to-silt size range. A good test for the likelihood of finding foraminifera in such "dirty" patches on the plant or animal surface is to swill the material vigorously under water and note whether any sediment remains adherent or not. If patches of sediment remain, then the likelihood is strong that foraminifera will be found in and under them, but if the surface appears clean and smooth to the naked eye, its yield will probably be slight. At this stage a hand lens is of value in judging the conditions of the substratum, but with a little experience one can operate effectively without it by this simple rule of thumb.

3. MINIATURIZATION OF EQUIPMENT

The conscientious biologist on a collecting trip should take seriously his responsibility of disturbing the environment as little as possible. His samples should be only large enough for his immediate needs and he should make an effort to avoid such excessive concentration on a small area as might initiate a concatenation of deleterious effects on the environment. With this in mind, attention should be given the miniaturization of equipment for the efficient collection and subsequent treatment of small samples with minimal impact on the area. Moreover, in these days of easy and rapid travel, particularly by air, a premium is often placed on small, lightweight equipment. Properly equipped for the task, the foraminiferologist can make valuable collections and work efficiently and speedily though remote from home base. Importantly too, this can be done with minimal expense and without the elaborate support commonly thought to be essential. [Arnold (1962) for example, has described the construction of an inexpensive and essentially expendable lightweight, high-speed plankton sampler suitable for manual operation onboard commercial vessels, freighters, pleasure craft or even ferries not equipped for oceanographic research or prepared to reduce speed to that required for towing conventional nets.] Modern air travel affords almost limitless access to collecting sites and to opportunities that are often merely coincidental to other activities, if only the logistics of collecting can be mastered. Miniaturization and modern plastics technology are the key to such mastery. Much of the equipment described here has been designed with this need in mind; a complete working kit has been designed to be carried in hand luggage by plane for use at the seashore and laboratories thousands of miles from my home laboratory. In this way, too, one can develop a portable laboratory that permits effective work at the seashore in improvised shelter, a motel, rooming house, or accommodation other than a marine laboratory when access to such a luxury is, for one reason or another, impossible.

In the miniaturization of field and laboratory equipment, as well as in innovative design generally, the feasibility of adapting some of the more sophisticated techniques for forming plastics to laboratory use should be emphasized. Vacuum-forming and compression molding, for example, though commercially performed on expensive equipment, can easily be accomplished on a small scale with inexpensive homemade equipment. These two processes alone open unimagined horizons for the fabrication of useful field and laboratory equipment. Small objects can be vacuum formed with the most rudimentary tools. A disc of suitable plastic (polyethylene, cellulose acetate butyrate, etc.) is clamped over the mouth of a small chamber from which the air can be evacuated (mild vacuum of high volume) as soon as the plastic is adequately softened by the heat of a Bunsen burner or electric heating element. (For small objects one's own lungs are an adequate source of vacuum, a match or candle an adequate source of heat). The inspection dish illustrated in Fig. 4b and the body of the mini-sieves shown in Fig. 12b were vacuum-formed; these are but two of several different items made for laboratory use. Match-molding of thermoplastic sheets is also a simple process, illustrated in Fig. 9a. The disc of plastic is clamped between two large metal washers, heated over a Bunsen burner, then quickly placed between the male element of a mold clamped in the drill press chuck and the female element clamped to the table of the press in a vice to assure proper alignment. While still soft the plastic is formed by closing

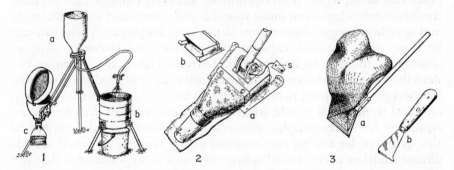

Fig. 1. Apparatus for concentrating foraminifera at the seashore. a. Tripod-mounted reservoir supplying seawater for the washing operation. b. Nested sieves on detachable legs, collecting bucket beneath. c. Bagging funnel with sieve in place, showing method of transferring concentrate from sieve to plastic bag under field conditions.

Fig. 2. a. Epibenthic dredge for hand use in shallow water, showing detachable net and glass vial, cotton apron, aluminium body and runners, detachable handle (three lengths, totalling 1.2 meters), and attached screwdriver (s) for altering angle of handle. b. Orifice view of aluminium housing. Lower lip of orifice can be bent to angle suitable for various substrata. See Fig. 13 for pattern of dredge housing. Dredge width: 10 cm.; housing length: 8 cm.

Fig. 3. a. Short-handled scraper for collecting algae bearing foraminifera. The bag should be of tough, closely knit material. b. Scraper for use in collecting sediment and organic material adherent to rock surfaces.

the mold upon it. The micro-centrifuge tubes shown in Figs. 6a, 7, 9b, and 11 were formed in this way, as were the coverglass dish in Fig. 9d and the inspection cup in 9c. The adaptation of extrusion and injection molding processes to laboratory use further extends one's design capability and need not be an expensive investment. Homemade apparatus for these operations has been illustrated by Arnold (1964, 1965), and Swanson's (1965) account of plastics technology is a valuable practical guide to more efficient equipment.

B. Collecting Tools

1. SCRAPERS AND KNIVES

In collecting those intertidal foraminifera found on rock surfaces or associated with attached plants and larger animals, scraping and cutting tools are of great value. A putty knife with a blade 3–4 cm. wide and 7–8 cm. long makes an efficient scraper if its tip, after heating, is bent downward through about 100 degrees and retempered to leave an inturned scraping surface 1–2 cm. high (Fig. 3b). A hunting knife with a sturdy blade and stout handle is most useful for prizing algal holdfasts from rock surfaces or for collecting algal fronds and marine grasses

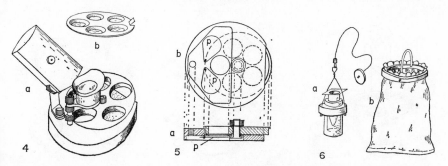

Fig. 4. Field microscope. a. A 10 × hand lens, shielded with plastic eye cup, can be focussed by means of thumb screws to permit viewing the individual compartments of the specimen tray (see b.) as each is rotated into position by means of the clear acrylic stage, illumination furnished by pocket flashlight when sunlight is inadequate. b. Specimen tray, vacuum-formed of cellulose acetate butyrate. The tray is aligned over the central knurled pivot of the stage; the upturned tab facilitates handling. See constructional details in Fig. 5.

Fig. 5. Constructional details for base of field microscope. a. Vertical section to show relation of rotating stage to base. The knurled pivot is machined to hold the stage firmly in place while permitting easy rotation. Dotted line in milled recess on left side of base indicates pyriform piece of plastic (see p in b.) used to provide suitable contrast for viewing by reflected illumination. b. Plan view of base from under side. Note tangential milling cut on right periphery of base which permits one-finger rotation of the stage.

Fig. 6. a. Sling centrifuge for field use. See Fig. 9a for method of forming plastic centrifuge tubes. Dimensions: height (less tube 18 mm., greatest diameter 11 mm., internal diameter 8 mm.. b. Collecting "bucket", consisting of heavy polyethylene bag held open by embroidery hoop. Handle is a short length of plastic-coated electrical wire.

suspected of harbouring foraminifera, but to avoid accidents while scrambling about over slippery intertidal surfaces it should be securely stowed when not in actual use. Two straps of polyethylene rivetted to the inside or outside of a plastic collecting bucket make a handy sheath for the knife, one superior, in fact, to the usual sheath, because the latter collects dirt and seawater which are difficult to remove and quickly abrade or corrode the blade unless the sheath is carefully cleaned after each use. The straps may easily be rivetted to the bucket by inserting short lengths of polyethylene rod stock into predrilled holes and flattening the exposed ends with a flat-tipped soldering iron. The putty-knife scraping tool may also conveniently be held by similar straps specifically designed for it and attached to the side of the bucket.

2. DREDGES

One of the simplest of the dredges for effectively sampling loose sediments suspected of containing living foraminifera is a short length (15 cm.) of iron pipe (5–7 cm. i.d.) closed at one end and towed by means of a

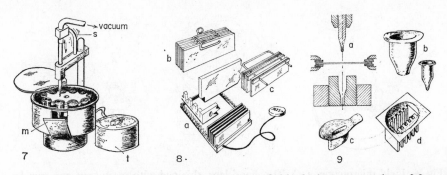

Fig. 7. Aspirator centrifuge for use in changing fluids during preparation of foraminifera for cytological study. Flexible support (s) of vacuum pipette permits lowering tip into centrifuge tubes, the operation observed by means of mirror (m) directed at viewing port in side of housing. Centrifuge speed controlled by toroidally-wound transformer (t). Overall height of centrifuge: 12 cm.; housing diameter: 38 cm.

Fig. 8. Harvest racks for foraminifera. a. Polyethylene tray with glass microscope slides. Lower corners of slides are notched to fit slots in tray. b. Harvest array consisting of microscope slides held by rectangular coil of stainless steel wire. c. Harvest array consisting of clear polystyrene strips threaded onto stainless steel wire frame. A float cut from foamed polystyrene (one attached to rack in a.) marks the position of racks planted in deep marine tanks or in the sea.

Fig. 9. a. Match-mold procedure for forming plastic products used in study of foraminifera. The two elements of the mold are closed onto a securely clamped thermoplastic sheet softened by heat. b. Vial-filling funnel (upper drawing) and microcentrifuge tube (lower drawing) formed in polyethylene by the process illustrated in 9a. Tube capacity: 0.3 ml. c. Inspection cup of clear plastic (cellulose acetate butyrate) formed by same process. The cup is used with a hand lens for the inspection of specimens in the field. Cup body length: 12 mm., height: 10 mm., capacity: 0.2 ml. d. Staining dish for coverglass preparations, prepared by process illustrated in 9a.

line attached (at one point only) along the leading edge of the opposite end. Because of its weight, the pipe digs into the sediment more effectively than lighter samplers and it requires no separate weights, depressing devices, or other sources of complication and difficulty.

A pipe dredge is useful for sampling from shore or seaward of one's wading limits, but the depth to which it penetrates the sediment is not easily controlled and it usually retrieves much of its load from the reducing environment beneath the layer of oxidized sediment at the water-sediment interface, so the percentage of dead specimens and empty tests thereby procured is usually high and the ratio of tests to sediment low. For this reason, a sled or epibenthic dredge is generally preferable for collecting living foraminifera. With a dredge designed to skim the sediment surface and dig only a few millimeters at most into it, a higher

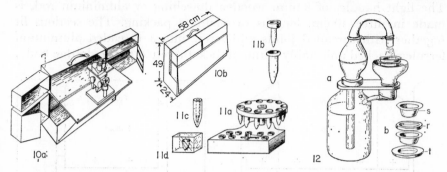

Fig. 10. Field desk for study of foraminifera and other micro-organisms. a. Desk in open position, with all compartments and work surfaces accessible. b. Desk in closed position, ready for carrying.

Fig. 11. Apparatus for agar embeddment of foraminifera prior to paraffin embeddment and microtomy. a. Loading and storage rack for agar-molding cores and microcentrifuge tubes (see b.). b. Microcentrifuge tubes (lower) and brass core (upper) for use in molding agar inserts into which foraminifera are concentrated by centrifugation. c. Agar insert (after extraction from centrifuge tube in which it was molded), showing cavity in which specimens are concentrated by centrifugation in apparatus shown in Fig. 7. d. Agar cube (greatly enlarged) cut from lower end of insert shown in c. Specimens are sealed in cavity by means of heated needle. The agar cube is then treated as a single piece of tissue during subsequent paraffin embeddment and microtomy.

Fig. 12. a. Apparatus for washing and sieving foraminifera from sediments or algal washings in the field or laboratory. Water, pumped by valved bulb through spray nozzle onto stack of vacuum-formed minisieves, is constantly recycled. b. Details of sieve stack, showing two of the four vacuum-formed minisieves (s), a polyethylene spacer (r) used between the sieves (to accelerate the passage of fluids through them), and the base plate (t) that supports the stack inside the washing basin. Overall height of apparatus: 30 cm., maximum width: 15 cm., minisieve diameter: 46 mm. (max.), height: 19 mm.

percentage of living specimens is obtained, the overall ratio of tests to sediment volume is generally more favorable, and yields following further concentration by sieving are generally superior.

A practical and versatile epibenthic dredge for certain types of intertidal and shallow-water studies is the small hand-operated tool shown in Fig. 2. Constructed of light-weight material (aluminium) and of such size as to make it portable and easy to pack for travelling, this dredge weighs only 184 grams (including handle and vial) and, except for the removeable handle, easily fits into a field-jacket pocket. A pattern for the body is given in Fig. 13. Nets of different mesh size may be attached to it by means of snap fasteners. Screw-cap glass vials are attached to the cod end of the net by means of a vial cap that has been bored on a lathe to permit the net to fit through it. The vial is then screwed into place in the cap, the net firmly clamped between the matching threads. The light handle, of 8 mm. wooden dowelling or aluminium rod, is made in three 40 cm. lengths to facilitate packing. The sections fit together with threaded joints and screw into a threaded aluminium ferrule attached adjustably within a socket rivetted to the dredge body.

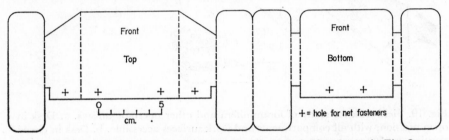

Fig. 13. Pattern for metal body (Fig. 2b) of epibenthic dredge shown in Fig. 2a.

A small piece of sturdy metal shaped to form an effective screw driver is attached by stout twine to the socket for use in adjusting the angle of the dredge. An apron of cotton twill should be attached to the body by the same snap fasteners that attach the net itself. This apron protects the net and vial during use. The lower lip of the dredge's orifice (Fig. 2b) can be bent downward or upward to regulate the depth of sediment penetration, making the device useful in a fairly wide range of environments. It is most effective when used on loose soft sediment in the range from mud and silt to fine sand, but coarser sands are easily sampled too, and rock surfaces, particularly smooth ones, may be scraped effectively if the lower lip is bent downward to make good contact and excessive pressure is avoided during use. The surfaces of marine grass blades or expansive masses of flat-fronded algae may be scraped with it whether they are submerged and upright or exposed and prone, but a tufty growth of plants or animals is best sampled with a conventional scraper (short-handled for hip-boot work in rocky areas, long-handled for deeper water). The dredge has also proved effective either as a scraper or as a true epibenthic sampler in collecting specimens from large aquaria, holding tanks and wet tables at marine laboratories.

C. Field Examination

1. HAND LENS

The ability to examine samples in the field and from this to evaluate the type and quantity of various species they are likely to yield when ultimately prepared for study can greatly increase the productivity of one's field effort. A simple hand lens of 10 to 20× magnification is the easiest instrument to use while one is actually wading about making collections; but, when used in the usual makeshift way, it also gives the least satisfactory results, the problem basically being that of achieving proper illumination and contrast, keeping the specimens immersed for best image quality, and having a sample of uniform particle size, so that small specimens are not obscured by large particles of sediment and debris.

Several different approaches immediately suggest themselves to the experimental field worker. By immersing the sample in a glass vial or plastic bag of clean water, the optical conditions for examining it are superior to those when studying it in air, but the rounded glass surface is a source of distortion and the plastic bags quickly cloud up unless kept wet on the outside. Small, shallow, flat-bottomed dishes vacuum-formed from transparent plastic (cellulose acetate butyrate) are useful if a black background can be provided when needed for proper contrast, but the little inspection cups illustrated in Fig. 9c have proved superior to any of these for general use in conjunction with the hand lens. The sample-containing well of the cup is 10–15 mm. deep and of similar length, but only 1–1.5 mm. wide, so that a good ratio between visible surface area and sample volume is established. The sample is easily examined through the clear wall of the well and the cup can be tilted at almost any angle to achieve proper illumination without the risk of losing the sample. The cupped top of the cell makes the introduction of the sample to the well by pipette, knife blade, or small spatula a simple matter. Such inspection cups are easily made by the technique of matched molds illustrated in Fig. 9a for the preparation of micro-centrifuge tubes. The cups are so simple and inexpensive to make that they are expendable, an added advantage under field conditions where accidents easily occur, particularly when cold water numbs the fingers.

Just as providing a suitable container for immersing the sample during inspection greatly enhances the quality of the image, so does the reduction of a sample to fractions having uniform particle size facilitate its inspection. This fractionation is easily achieved through the use of a set of miniature sieves like the one described under "Preliminary Concentration" and illustrated in Fig. 12b.

2. FIELD MICROSCOPE

The logical outcome of the use of the various inspection devices described above was the development of the field microscope illustrated in Figs. 4 and 5. This is one approach to combining in one unit the major features required for the effective microscopic examination of samples in the field.

The microscope consists of: (a) a 10× hand lens equipped with an eye shield and attached to an upright along which it can be moved by means of knurled nuts for focussing; (b) a plastic base with rotating stage having five light wells; (c) a transparent, five-compartment inspection dish (Fig. 4b) that fits over the stage and rotates as a unit with it; (d) two pyriform pieces of opaque plastic (one white, the other black) mounted in a milled recess beneath the base and pivotted so as to be used interchangeably to provide a background of suitable contrast for viewing by reflected illumination; and (e) a pocket flashlight (a source of illumination for use when sunlight is inadequate) so mounted as to be easily adjusted when needed or removed when not.

With this microscope it is possible to examine a sample much more critically than with a simple hand lens. It has even proved practicable in insolating individual foraminifera through the use of a mouth pipette while still in the field, but, of course, its use does require more time, more preparation before use, greater care and cleanliness during use, and more careful cleaning after use than does the hand lens alone, so one must evaluate the two for himself under actual field conditions.

The eye shield over the lens is a great advantage in the field and is of critical importance to efficient examination. By removing the lens housing and its pivotting mount from the safety cover into which they normally fit, and by turning down the housing slightly on a lathe, a plastic lens cup can be fitted snugly and securely to it. Focussing is accomplished by means of two knurled nuts that imprison the lens mount and move it along the threaded upper half of a brass upright screwed into the base of the instrument. The clear acrylic (plastic) stage rotates in a recess milled into the under surface of the base. The edge of the stage is exposed in a flattened section ground from the base's periphery opposite the lens stand; this facilitates rotation by means of the viewer's finger. Each light well in the stage can be rotated in turn over a corresponding hole in the base which gives access either to transmitted light or to the pyriform plastic pieces that provide suitable background contrast for reflected illumination.

The clear, plastic inspection tray simply slips over the central shaft of the stage and rests upon the latter, the five depressions in the dish nestling in the corresponding holes in the stage. A small upturned tab along the periphery of the dish facilitates emplacement and removal. The depressions in the dish should be of sufficient depth to permit immersing samples of sand-sized particles, *i.e.*, at least 2 mm. at the deepest point. To facilitate examination and subsequent cleaning, the sides of the depression should taper and the bottom should be flat. The two pyriform pieces of plastic that serve as interchangeable backgrounds should be sufficiently thick for easy adjustment by finger. They are held in place by and pivotted around small screws threaded into the recessed segment of the base within which they swing. The pocket flashlight is held to the base of the microscope by means of a knurled screw and a simple bracket of flat brass bent to suitable shape and attached to the lamp housing by a screw which passes between the parallel batteries within the case. Both the screw on the base and the one on the lamp housing can be loosened to permit adjustment of the light and then retightened for stability. The instrument here illustrated is for use with the left eye and to be held in the arc of the left thumb and forefinger, so that the right hand can be used in manipulating the inspection dish, the mouth pipette and dissecting needle or other tool.

3. FIELD DESK

Though it is generally preferable to return to an established laboratory for the final preparation and study of collections, a portable field desk of the type illustrated in Fig. 10 can often greatly increase the productivity of one's field effort. The author's has repeatedly proved its worth, particularly when collections were being made in areas remote from a laboratory or when instruction was being given students in the field. Containing all the equipment for concentrating, cleaning, examining and killing-fixing specimens, the field desk is truly a portable laboratory that can provide all the facilities one needs for the effective treatment of samples. It can change a haphazard effort to a highly efficient one. The field desk has been useful in the summer months when most marine laboratories are particularly crowded, because one can operate with it in the quiet of the field or in a nearby motel or rooming house, only using the laboratory for especially elaborate or otherwise demanding needs.

In use the folding desk expands to provide a total of 0.37 square meters of surface at or near the level of the microscope itself. The folding compartments afford ample storage space for microscope, light, sieves, reagents, and the various tools necessary in the preparation and study of the samples. The unit folds compactly for carrying (Fig. 10b) and, though of light construction, is sturdy enough to give adequate protection to the gear it contains.

D. Preliminary Concentration (Field)

Unless suitable laboratory facilities are available, it is often easier to wash and concentrate samples in the field where an abundant supply of seawater is at hand and surplus materials can be returned directly to the sea. When working in shallow water over a gently sloping sea floor, the use of lightweight sieves that can be carried and used while wading saves much time spent walking the long distance from the collecting site back to shore. Sieves for this purpose can readily be constructed by cutting the lower third from nesting polyethylene basins (using a fine-toothed circular saw or a heated resistance wire) and welding metal screening or nylon netting to a rim left in the bottom of the basin when the opening is cut into it. A soldering iron equipped with a spatula-like blade is an effective tool for the welding operation, particularly if a thin sheet of "Teflon" is placed between the tool and the strips of welding plastic. The sieves so produced then fit into a similar basin left intact to serve as a collecting pan for each fraction in turn as the various sieves are worked. By rivetting and welding strips of polyethylene within the basin to form an internal pouring channel, the collected residue can easily be transferred to a bagging funnel without loss and thence to a polyethylene bag for return to the laboratory. Shoulder straps of macramé hold the basin at a convenient working height, and the basin serves as a support for the individual or nested sieves during the actual washing. To complete the unit, a bagging funnel attached to one's belt or supported by a rod that can be stuck into the sand facilitates the final transfer of sieved fractions to plastic bags. The funnel is made by cutting the upper half from a plastic bottle having an integrally molded handle, which is easily adapted to fit one's belt or the top of a support rod. The original cap of the bottle should be bored out on a lathe for use as a threaded ring to hold the plastic bag over the narrow orifice of the funnel.

With this equipment it is possible for a wader to wash sediment, algae, grasses or other materials suspected of harbouring foraminifera, to concentrate the specimens effectively, and to examine the concentrate with a hand lens while still in the water and some distance from shore, often saving valuable field time by not having to walk back to shore for these operations. The efficiency of one's field effort can be greatly enhanced and a higher level of environmental and faunal analysis achieved through such on-the-spot preparation and evaluation rather than through the haphazard and essentially blind collection of samples neither concentrated nor adequately evaluated in the field. Like much of the equipment described in this account, its light weight and compactness make it particularly well suited to collecting trips remote from one's home laboratory, especially when air travel, with its restrictions on weight and space, is involved.

For intertidal collecting or for collecting in waters over bottom slopes so steep and narrow that one must work either on or very close to shore, the apparatus shown in Fig. 1 is preferable. Though relatively heavy and designed specifically to be carried by automobile, this apparatus is still portable and can readily be handled by one person, even in the hazards of an algal-rich rocky intertidal area. A stream of seawater from an overhead reservoir (one-gallon polyethylene bottle, upended, its bottom first removed) attached on top of a sturdy photographic tripod can be directed into a set of nested sieves supported on a short-legged frame. In this case the bagging funnel is attached to the legs of the reservoir tripod, but it is used just as described above. Most of the washing should be done by first scrubbing the algae on a coarse sieve in a basin and then pouring buckets of water onto the concentrate on each of the nested sieves in succession, but the final washing can be done from the overhead reservoir; it is particularly useful in swilling the washed concentrate from sieve to bagging funnel.

A device which has proved of great value in the preparation of small samples for examination with hand lens or field microscope in the field or regular microscope in the laboratory is the miniaturized washing apparatus illustrated in Fig. 12. Merely the means for directing a strong steam of recycled water through a nesting set of miniature sieves, the device facilitates the efficient washing and concentration of either living specimens or their empty tests. Water, pumped from reservoir to screens by means of a valved bulb, is recycled automatically so that the unit is particularly advantageous in inland situations or wherever sea water must be economized, but the convenience of having such complete washing facilities just at one's microscope is appreciable even in the best-equipped of marine laboratories. Small and light, it is also well adapted for travel and for use at the seaside, on shipboard, or in laboratories remote from one's permanent preparation facilities. It provides the means for rapidly extracting and concentrating foraminifera in minimal volume and in a viable state for subsequent treatment.

The apparatus, 30 cm. in height and 15 cm. in maximum width, consists of a polyethylene reservoir (bottle with a working capacity of one litre), a valved bulb (65 ml. capacity) fitted with rubber intake and delivery tubes (the latter having a plastic nozzle designed like a miniature shower head), a polyethylene basin (cut from the top of another polyethylene bottle) in which the mini-sieves are nested and from which a return tube directs the water back into the reservoir. The nozzle of the delivery tube is held in position over the stack of sieves by a length of wire attached to the upper end of the bulb. The pattern of holes drilled in it (#70 drill) and the angle at which the outer rows of these are drilled in the nozzle should be such that its spray covers the entire floor of the sieve. One hand is used to operate the bulb, the other to control the nozzle and agitate the screens to accelerate the washing process.

The mini-sieves (s, Fig. 12b) used in the apparatus have a rim diameter of 5 cm., a base of 3 cm. (o.d.) and a height of 18 mm. The body of these is vacuum-formed from polyethylene sheet (1 mm. thick), and discs of metal or nylon screening are welded firmly in place on an internal rim by means of heat and pressure applied through simple molds operated in a drill press. Polyethylene spacers (r, Fig. 12b) between the sieves are desirable for efficient operation of the stack, the entire assembly held in place within the basin by means of a polyethylene disc (t, Fig. 12b) cut to fit the contours of basin and sieves.

Though its weight (216 gms.) is no problem, should the entire unit prove too cumbersome for the wader, the bottle/basin assembly can be left behind, the bulb unit and sieves easily carried in a pocket or field apron ready for instant use. The sea then serves as the reservoir; the sieves are held in one hand, the bulb operated by the other. In this case, however, a fine screen should be placed in or over the mouth of the intake tube to prevent particles suspended in the water from clogging the valves or nozzle. A small vacuum-formed plastic dish can be used as an examination dish in conjunction with the sieves, a part of each sieved fraction being transferred directly to it from each sieve in turn for examination with hand lens or field microscope. The spray from the nozzle can be used quite efficiently to effect the transfer.

E. Transporting to Laboratory

Most materials from which foraminifera are subsequently to be removed in the laboratory are readily transported in an open bucket or plastic bags, the principal precaution being to pour off all water while keeping the material sufficiently moist to sustain microscopic life. To prevent actual desiccation, algae or loose wads of moistened newspaper may be interspersed with the material and spread lightly over the top. It is essential to provide maximum contact with air, but humid conditions must be maintained in the container itself. Water held in the interstices of the materials from the sea and as a thin film over their surfaces is generally adequate to prevent damage, if the precaution of insulating the whole with a loose wet cover is respected. Leaky plastic bags and buckets are preferable to water-tight ones for the same reason, the risk of suffocation as the oxygen in stagnant water is exhausted being particularly serious. Where possible the containers should be left open at the top, but if they must be closed to prevent spillage, or if the journey is sufficiently brief, then commercial bag ties or ties made by cutting plastic-coated electronic wire into short lengths may be used. For small samples, plastic or glass vials are useful.

A method that has proved satisfactory in transporting sediment concentrates as well as algal mats containing foraminifera—particularly when travelling long distances by air or when the material must be kept *en route* for a few to several days without refrigeration and with minimal attention—is to spread each sample thinly onto pre-cut pieces of moderately absorbent paper that can then be stacked and kept well moistened—but not water-logged—in small plastic bags. Water can readily be drained from the bag to prevent the suffocation that would otherwise so quickly follow if it were left stagnant in the paper. Periodically the stack can be remoistened and drained, it being unnecessary to remove it from the bag for the operation. A much greater volume of sample material and a larger number of samples isolated from one another (to avoid contamination) can safely be transported in this way than in vials occupying the same overall volume, and the hazards of fouling or break-age are far less. When one's destination is reached, each sheet is individu-ally swilled into a separate dish of seawater, or alternatively, it can be immersed in the dish and shaken or brushed lightly to dislodge the specimens, the paper being removed subsequently. Care should be exercised in the choice of paper used for this purpose: chemical contami-nation should be avoided, as should excessively coarse or loose paper that would swell or rapidly disintegrate when wet. The surface should be reasonably hard so that the specimens do not become entrapped in the fibres of paper, but it should be sufficiently absorbent to hold enough moisture to prevent rapid desiccation.

F. Miscellany

Some simple means of concentrating specimens for killing and fixation either in the field or wherever electrical or chemical centrifuges are not available is often desirable. For this purpose the sling centrifuge illu-strated in Fig. 6a has proved useful. The cup, turned from acrylic stock and polished to transparency, receives small (0.3 ml. capacity) centrifuge tubes compression molded from low-density polyethylene sheet stock, the forming process illustrated in Fig. 9d and described elsewhere in this article. To use the centrifuge, it is merely whirled rapidly by hand from a length of twine.

The superior qualities of plastic buckets as general-utility containers for intertidal and shallow-water work have long been recognized by marine biologists, but a practical substitute for use by the air traveller who must conserve luggage space or for anyone else who requires an extremely light collapsible container is the improvized bucket illustrated in Fig. 6b. Consisting merely of a polyethylene bag held open by inter-locking embroidery hoops to which a wire or rope handle has been fitted, this "bucket" can easily be carried in one's personal luggage onto a plane or kept in a field-jacket pocket ready for instant use.

A handy holder for the various tools one uses while in the field can be made from a carpenter's apron to which additional pockets are sewn. Such a compartmentalized apron keeps the tools readily available and secure from loss, so that it is an easy matter to examine samples, con-centrate them, and even isolate specimens of interest before leaving the collecting site. Time spent in designing pockets for the efficient accommo-dation of the various tools and supplies one requires while at the seashore pays rich dividends under conditions not always the most favorable for the demanding type of observation necessary in the careful field study of micro-organisms.

IV. Pre-Culture Maintenance and Preparation

A. Interim Care

Immediately on return to the laboratory, steps should be taken to prevent the rapid fouling or deterioration that naturally occurs in freshly-collected materials left without proper attention. If running seawater or established aquaria are available, particularly ones of large volume, the material may be placed in these, spread as thinly and kept as dilute as the available facilities allow.

The samples should be placed in holding tanks, trays, or aquaria supplied with a good flow, but care should be taken to avoid overcrowding, and, in general, the same procedures should be followed as those used in dealing with larger invertebrates and algae at this "holding" stage. Algae awaiting scrubbing for the release of foraminifera should be kept vigorously swirling in the water, and sediment awaiting treatment should be allowed to form only thin layers (a few millimeters thick at most), the finer the sediment the thinner the layer.

In the absence of running seawater or proper recirculating facilities, however, all material should be spread out in shallow trays with nothing more than interstitial water and a covering layer of dampened paper towelling to keep them moist until the foraminifera can be prepared for harvesting and concentration, the idea here being to simulate conditions encountered by the foraminifera when left amongst the algae exposed by receding tides. If some means—such as a moist chamber or a large, closed plastic box—is available for creating and maintaining a humid environment for materials from which all supernatant water has been drained, the foraminifera stand a far better chance of surviving than if they are left submerged with large quantities of organic debris to foster bacterial growth and other changes that rapidly prove disastrous for them. A wise precaution to exercise until proper cultures can be established is to lower the ambient temperature to or near that of the winter temperature of the water in the collecting area. This reduces the rate at which oxygen is depleted, retards bacterial growth and the decay consequent upon this, reduces evaporation, and gives one more time in which to deal carefully with the samples.

B. Examination and Evaluation of Field Collections

1. GENERAL CONDITION OF SAMPLES AND SPECIMENS

The direct microscopic examination of living and non-living material suspected of harbouring foraminifera can enhance one's understanding of the organisms' bionomics and general biology, can lead to the discovery of delicate and small forms not recovered by or relationships lost during the usual concentrating and harvesting procedures, and can enhance one's efficiency in subsequent collecting and field study. The different types of material on or in which foraminifera live (algae, grasses, invertebrates, rock, and sediment) require slightly different techniques of preparation and conditions for examination, the basic requirements for efficient examination being those already enumerated for microscopic examination in the field.

Using a stereoscopic loupe mounted on the forehead, and guided by accumulations of silt-sized particles that remain on the surface of the material after vigorous swilling in an aquarium, pan or bucket of sea-water, the more likely materials from the first four types of substratum can readily be sorted from the barren ones for further consideration.

The likely areas of algal fronds should be cut from the main mass with scissors or other suitable tool, the individual pieces rendered sufficiently small to fit a shallow container in which they can be flattened and totally immersed. The sides of the container should be sufficiently high to prevent spillage when handled with reasonable care and speed. Algal holdfasts should be cut or torn apart and reduced to pieces sufficiently small for the effective examination of all exterior surfaces, and the numerous nooks and crannies should be bared for careful scrutiny. Haptera that have been opened to the elements by decay and erosion should be slit lengthwise and spread out to reveal the miliolids and opthalmidiids that appear to have a predilection for this habitat.

Because of their simple shape and relative uniformity, the blades of marine grasses lend themselves to techniques of rapid preparation and examination. Large quantities of the more promising segments of "dirty" blades can be cut into short lengths convenient for handling in the inspection tray; each side of the blade should then be examined with light striking at a low angle from above.

Aborescent or tufty invertebrates should be cut or teased into pieces sufficiently small and flat for the complete examination of each constituent branch or element, the principal desideratum being that the overall thickness, or height, of each piece be so reduced that frequent focal adjustment is obviated. [A foot-controlled focussing device for the dissecting microscope should be useful in such operations as this. An effective one has been described by Strong (1969); commercial models are available.] Small fleshy invertebrates can be examined *in toto*, but larger ones should be so cut that their external surfaces can be immersed. The shells of invertebrates can similarly be examined completely if small, or broken or cut (hacksaw) if large. Both the inner and outer surfaces of shells can be a rich source of attached foraminifera, while the interior of decomposing or weathered shells might harbour such interstitial forms as *Shepheardella*, revealed on delamination.

Pebbles and small fragments of rocks can be examined without additional preparation, but larger rocks must be reduced with chisel and hammer, heavy pliers or cutters, or more powerful tools, such as diamond saws and hydraulic shale splitters. Only the external surfaces of the harder rocks need generally be examined, but all the surfaces of well-weathered rocks, especially such fissile ones as shales, should be examined following delamination, particularly when evidence of incipient delamination through natural agencies can be detected.

Unless special efforts were made in the field to obtain sediment samples in an undisturbed state, the microscopic examination of raw sediment prior to separation into size classes through sieving is not only an exceedingly tedious and inefficient process but one of questionable value as well. Because loose sediment generally affords less protection from abrasion and predation for the more delicate species of foraminifera (allogromiids, for example) than do fissile rocks or the interstices of plants and colonial invertebrates, the yield from it following sieving more nearly reflects the natural population than does that following the scrubbing and sieving of material from these other sources, substrata from which the more delicate specimens might be destroyed during preparation.

The tools of use in baring specimens to view and then removing them from the materials on or in which they occur include fine brushes (for removing the sediment or debris that often covers attached forms), fine entomological pins (*minuten Nadeln*) equipped with applicator-stick handles, dental tools (ground or otherwise reshaped and then sharpened to form miniature chisels and scalpels), and mouth pipettes. The early development of ambidexterity in the use of these tools under the microscope pays rich dividends for the serious microscopist who wishes to study foraminifera.

2. DISTINGUISHING LIVING FROM DEAD SPECIMENS

Before investing time and effort in concentrating or isolating specimens and in establishing cultures with them, it is important to distinguish living from dead individuals and to obtain some idea of the relative proportion of the two in the samples, so that particular attention can be given the more promising ones. Aids in making the distinction are also important after the cultures have been established and whenever their condition must be evaluated; techniques toward this end are essential to the successful conduct of the culture program.

Anyone attempting to make the distinction within a diverse array of species, however, quickly becomes impressed with its difficulty. Although considerable effort has been devoted to the search for a simple and universally effective means of distinguishing between the two, none has been found. Some of the numerous pitfalls awaiting the user of the rose Bengal technique developed by Walton (1952) and widely employed by foraminiferan ecologists and paleoecologists subsequently have been pointed out by Le Calvez (1972). She oversimplified the difficulties when she stated that the only species that cannot be distinguished simply by direct observation are the arenaceous ones. Her contention that all calcareous forms have such a distinctive color in the living state as to be readily distinguishable from dead ones fails to consider the numerous real difficulties with which foraminiferal biologists have long been grappling as they attempt to evaluate both natural and laboratory populations of these organisms. While direct observation is adequate at all stages in the life of some species or at a certain stage—principally the youthful one—in the development of many, the problem can be difficult indeed in others, such as those having heavy, opaque tests, those in which large quantities of organic debris remain in or are washed into the test following reproduction or after death, those that remain attached to the substratum after reproduction or death, those capable of retracting their intrathalamous protoplasm into a relatively small mass within localized parts of the test lumen and remaining dormant for protracted periods of time, those having large numbers of symbiotic plant protists within their tests, and those invaded by other organisms.

While a single simple and foolproof technique for making the distinction in all species and at all periods of their life has not been encountered, a wide assortment of indications and tests is available which together will cover practically any species or circumstance, the problem then being to select the method or combination of methods best suited to the particular situation. Basically, the clues or tests derive principally from: (1) movement of the entire organism, including observed autonomous movement as well as that inferred from the displacement of the test in the absence of external forces, and that clearly implied by paths left on or through the substratum by a moving specimen; (2) protoplasmic movement (pseudopodial streaming or intrathalamous cyclosis) and the direct consequences of this, including the development of pseudopodial tracks and tunnels; (3) feeding activity and the direct consequences of this, including the accumulation of food clumps and debris in the region of the aperture or over other parts of the test, and the development of protoplasmic color (green particularly) from the ingestion of food particles; (4) test sheen; (5) *in toto* staining (vital dyes, rose Bengal); (6) cytological study; and (7) crushing the test to reveal the condition of the protoplasm.

Ideally, the method employed should in no way harm the animal or interfere with its normal activity, unless enough specimens are available to permit such sacrifice; but, as will shortly be seen, a compromise often has to be made. It is often easier, for example, to detect pseudopodial activity with the higher magnification afforded by a standard monoscopic microscope than with a stereoscopic instrument, but if the specimen happens to be well "bedded down" in an important experimental culture and in no condition to be removed for examination on a microscope slide, then means for distinguishing with the latter instrument must be devised. At such times, then, it is well to have different techniques and approaches at hand.

With most free-living animals, movement of the creature as a whole or of its parts is generally the most obvious evidence of life, but because most foraminifera move quite slowly, many remain immobile but quite healthy while feeding, growing and reproducing, and some attach themselves to the substratum early in life and remain attached thereafter, this criterion is of limited value. Applied with discretion, however, it can be a useful clue indeed. For example, if sediment concentrate is scattered thinly over the bottom of a shallow container through which seawater can flow and left overnight, many of the foraminifera contained within it will have crawled by the next morning to the surface of the sediment, up the sides of the container, or up any object that projects even slightly above the general level of the sediment. This behaviour has been exploited in the development of some of the harvesting procedures described herein. One need not, however, wait so long to detect movement in many species normally encountered in intertidal and shallow subtidal collections. Sediment similarly spread and examined within a few minutes under the higher magnification of a good stereoscopic microscope often contains living miliolids that are easily seen, the erratic movement of their tests reminiscent, in a lilliputian way, of that of snails. Careful microscopy, including resort to the higher magnifications of a monoscopic compound microscope, often reveals movement of other species as well, but, as indicated previously, serious limitations to the method are interposed by the very unco-operative nature of less active and sessile species. Viability and movement are simply not synonymous, nor are immobility and death.

If one takes the trouble to note the position of individual specimens at a given time and rigorously excludes any external factor (copepods, for example) that might displace them in the interim, any displacement that subsequently occurs can safely be attributed to autonomous movement and used as an indication of viability. Movement, and hence viability, can also be inferred from the paths left by the test as the animal grazes its way through its algal pastureland. Some species remain essentially stationary once they settle down in a well-fed culture, and others move so lightly about in feeding as to leave no discernible spoor, but several of the more successful laboratory species (the rosalinids, for example) leave easily followed paths as they scour diatoms and other food materials from the culture floor, pushing them aside and often burying themselves in the process. Well known to earlier students of the group, and described in some detail by Jepps (1942) in *Elphidium crispum*, the number of feeding cysts formed daily by experimental specimens was used by Murray (1963) as an indication of their "metabolic rate and healthiness".

The presence of pseudopodia, detected with either type of microscope is, of course, a clear indication of life, but the inexperienced observer should make certain that protoplasmic streaming is indeed taking place within any threadlike or filamentous extension from the test he encounters, lest he be misled by filamentous algae and other foreign matter that might superficially resemble pseudopodia at low magnification. Through the skillful use of light and contrast, pseudopodia can often be revealed in surprisingly small specimens under the stereoscopic microscope [Myers (1945), noted that they can be seen with a hand lens under field conditions.] Higher magnifications, phase-contrast, and dark-field microscopy are useful adjuncts in those cases where the specimen can safely be removed and subjected to the rigours normally associated with these more revealing procedures and when the results justify the additional time they require. The development of pseudopodia in otherwise inactive animals can often be induced by feeding, particularly if, in the case of non-circulating systems, the fluid is changed just before feeding. An hour or so later the pseudopodia, clearly indicated by the masses of food material adhering to them, can be seen radiating in all directions from the test, particularly if the culture floor has first been swept clean around the specimen. Some species, such as *Cibicides lobatulus* and *Planorbulina mediterranensis*, develop pseudopodial "tunnels" on the blades of marine grass (*Posidonia*) in the Mediterranean and are able to maintain these for long periods. The same process can be observed in laboratory populations of these two species. When brought into the laboratory the repeated extension of the pseudopodia into the tunnels and their subsequent withdrawal from them can be observed as food is added and the supply gradually exhausted. The accumulation of fresh masses of food around the open ends of the tunnels can easily be observed, its depletion followed over a period of days. Though few of the more successfully domesticated species develop such persistent tunnels in the laboratory, the temporary extension of pseudopodia, mentioned above, constitutes a similar, though ephemeral, indication of viability. The principal obstacle Nature has interposed against the universal reliance on pseudopodial activity is that many species may remain pseudopodialess for protracted periods and some cannot be enticed, even by the most imaginative and tempting of artifices, to produce them.

Inasmuch as the feeding process is generally dependent upon pseudopodial activity, other changes consequent upon feeding may clearly indicate that an animal is viable. These include the accumulation of masses of food material, sediment, and debris around the aperture or over parts or all of the test, and the color changes through which the protoplasm passes as ingestion and digestion proceed. One of the simplest ways of distinguishing between living and dead specimens is to observe the accumulation of food material and watch the changes in protoplasmic color that follow.

If before feeding (and changing the culture fluid), a few specimens are carefully brushed clean of adherent particulate matter, particularly around their apertures, the subsequent accumulation of food can be unequivocally noted and the temporary development of a green color by the protoplasmic body can be observed as ingestion proceeds. The food chosen for this test must, of course, be one that the species finds both acceptable and desirable. While it is reasonable and safe to conclude that the appearance of food material around apertures previously devoid of them indicates viability, it is not equally safe to conclude that the persistence of feeding clumps indicates continued viability, because, unfortunately, such accumulations do not necessarily disappear when the animal itself reproduces or succumbs to some intrathalamous parasite or predator. In many species the newly produced young derived from such an individual use the accumulations gathered by the parent as an immediate source of food and may remain in or on it until it is exhausted, but in others the young (or the gametes) rapidly move away from the parent, leaving the empty test buried in the mass. A covered specimen from which no pseudopodia are being extended may be only temporarily inactive rather than dead. Brushing away the food and debris may uncover a test sufficiently thin to reveal colored protoplasm inside, but this is not a sure indication of viability, because color often remains after the animal itself is moribund. The difference is essentially one of shade, darkening of the typical green, orange, or brownish-grey being the clue to decline. The change, however, is often so slight or subtle as to be unreliable; even such a dubious clue as this is of no value with specimens having opaque tests.

An indication of the general condition of an organism that can be observed rapidly and without moving or otherwise disturbing the animal (except possibly to brush away some of its algal cover) is the sheen of its test. A healthy specimen characteristically has a test from which light is reflected rather brilliantly when suitable viewing conditions are established. The "sheen of health" indicates generally that the test surface is clear of bacteria or the bacterially-produced film that develops soon after a foraminifer's decline sets in or death occurs.

Brushing away the accumulations of food and other particulate matter, especially when this is done a few to several hours after feeding or during the reproductive process, often reveals other evidence of viability: small masses of extrathalamous protoplasm may be seen around the aperture or around the periphery of the test when it is in contact with the substratum, and the extrusion of a protoplasmic lake, the frequent harbinger of asexual reproduction, can often be more easily detected when the "reproductive cyst", food or debris cover is removed.

Because of the numerous cytoplasmic inclusions in foraminiferal protoplasm and the difficulty of seeing through the test wall in many species, vital stains have not been extensively used in distinguishing living from dead specimens, but the rose Bengal technique referred to above, is widely used in spite of its recognized limitations.

If the circumstances of one's study permit the sacrifice of individuals, then two additional approaches may be employed in distinguishing living from dead specimens. The first of these involves the usual laborious preparation and study required for a cytological examination of nuclei and protoplasm, from which the state of the organism can be determined with a high degree of accuracy if the necessary baselines for that species have previously been established (a not inconsiderable limitation in view of the relatively few species that are cytologically well known). This method, however, is hardly practical. A second method, merely involving the crushing of the specimen to reveal the condition of its protoplasmic body, definitely merits consideration when specimens are sufficient to permit such sacrifices. Viable protoplasm is generally "sticky", readily adhering to the needles or other tools used in tearing the test open. Examined under high magnification, it has all the attributes of living matter, including residual cyclosis, and its components can be recognized by a competent cytologist and readily distinguished from the masses of bacteria normally associated with decay and death, or from the organic debris so often left inside the test following reproduction.

C. Concentrating Specimens

While it is not only feasible and practicable but often desirable as well to undertake some of the preliminary concentration of specimens before leaving the seashore, further concentration in the laboratory can usually be achieved through the use of the screening-washing procedures already described, by panning, decanting and flotation, pipetting, centrifugation, and through the use of harvest devices that exploit the natural tendency of vagile species to move up any vertical or inclined surfaces accessible to them. Whether the initial concentration was achieved in the field or in the laboratory, however, it soon becomes essential to separate the foraminifera from: (1) all organic debris that can foul the environment through decay; (2) any living organisms that constitute a potential threat either as predators and parasites or as poisoners of the environment through their own metabolic activity or following their death, and (3) all inorganic residue.

The fact that sieved concentrates generally contain materials of different specific gravity means that the biologist can profitably adapt the gold miner's and the economic mineralogist's techniques of panning, decanting and flotation to the further concentration of the foraminifera, a technique employed by Dujardin (1835, p. 346) in collecting his specimens. Each size fraction in turn can be placed in a shallow dish of seawater, where it can be swirled and agitated in such a way that the lighter materials are placed in suspension long enough to be poured off the heavier ones. If, through the use of a large pipette or vibration (mechanical or sonic, the latter used with great care to avoid damage to delicate specimens), the foraminifera are first loosed from their pseudo-podial attachment to other particulate matter in the sieved fraction, the ones with mineralized tests can often be concentrated beneath the bulk of the organic debris but above the heavier mineral matter comprising the inorganic sediment. Most of the lightest material can then be poured off, and a large percentage of the foraminifera can be removed from the sediment by repeating the panning operation. Needless to say, the method is not perfect, and resort must eventually be had to a pipette if all speci-mens are to be retrieved, but this simple procedure can eliminate some of the tedious sorting under the microscope. If tests alone are being sought, heavy liquids, such as those routinely employed by mineralogists and micropaleontologists, can be employed to achieve separation through differential flotation, but these, being poisonous, are not suitable for use with living foraminifera.

The ultimate stage of concentration requires the isolation of specimens from all unwanted material. For this, pipettes are the instrument of choice. A mouth pipette is probably the most efficient and precise tool for the purpose, although some workers prefer those operated by a bulb. If samples have first been separated into size fractions by sieving, the pipette used for isolating the foraminifera should have a tip just large enough to accept the largest specimens in that fraction, a different pipette being employed for each fraction. Preliminary sieving pays rich dividends, in the form of cleaner concentrates and more rapid harvests, at the pipetting stage. Before settling on a particular type of pipette, the beginner should experiment with more elaborate designs, including those with saliva traps and specimen reservoirs (Myers, 1933) and those operated by vacuum, a subaqueous modification of the sorting tools used by some micropaleontologists (Stinemeyer, 1965).

The value of a centrifuge in the laboratory is well known to the proto-zoan cytologist, and most commercial instruments can be used in carrying foraminifera through the various changes of fluid essential in the prepara-tion for cytological studies, but the device illustrated in Fig. 7 was designed specifically for this type of operation and the efficient handling of the relatively small volume of material with which the foraminiferal biologist is often confronted. The tubes it employs are the same as those used in the sling centrifuge, but glass ones are, of course, entirely suitable if they can be readily made or inexpensively purchased. The speed of the machine is controlled by a variable transformer. By means of its flexible plastic support, a vacuum-operated pipette can be pressed into each tube in succession for the removal of fluids, the operation viewed and controlled with precision through a mirror mounted opposite a plastic window let into the side of the centrifuge housing. A protective cover of clear acrylic can be pivoted into position when the motor is in operation but slides aside to permit free access to the tubes. When used in conjunction with extractible agar inserts in the centrifuge tubes within which specimens can be concentrated, the centrifuge can be of great value in preparing specimens for microtomy. Figure 11b illustrates a tube and the brass core used to mold an agar insert (Fig. 11c) in which specimens are concentrated by centrifugation. The insert is then removed from the tube and trimmed to a 2–3 mm. cube (Fig. 11d) in which the specimens are securely imprisoned by sealing the opening with molten agar; an electrically-heated needle is used to achieve proper fusion between the block matrix and the sealing agar. This block is then handled like a single piece of tissue in dehydration and further preparation for paraffin embeddment and sectioning.

The habit many foraminifera have of crawling up any vertical or inclined surface available to them can be exploited to separate living specimens from dead ones and to concentrate specimens from raw samples or sieved fractions of sediment or algal scrubbings. This is accomplished merely by setting racks or trays of microscope slides directly on top of material immersed in seawater and leaving them until the foraminifera have crawled from the sediment onto the slides. Inasmuch as the raw samples and sieve fractions generally contain large amounts of organic matter, circulating seawater and a slight reduction of temperature from that of the natural environment are particularly desirable during the harvesting operation if rapid fouling and the consequent reduction in the yield of living foraminifera are to be avoided.

Three types of array have been successfully employed for this purpose, two with standard microscope slides (Figs. 8a and 8b) and a third with plastic strips of the same size (Fig. 8c). Using any of the three, the foraminifera literally harvest themselves while one is otherwise engaged about the laboratory. The lowermost of the three arrays consists of glass slides mounted in rectangular racks made by welding appropriately slotted polyethylene strips to plain end members. (Welding is accomplished by means of a flat blade kept at low temperature in a rheostat-controlled soldering iron, the work itself held in a simple assembly jig to which mild clamping pressure can be applied when adjoining plastic surfaces are molten.) The microscope slides, a notch ground in the two lower corners of their long side, fit snugly into slots previously sawn in the lateral members of the rack by means of a thin-bladed circular saw. Figure 8b illustrates glass slides held by a rectangular coil of corrosion-resistant stainless steel wire, one end of which is bent into a loop by means of which the unit can be handled and the addition or removal of slides facilitated. The third type of unit (Fig. 8c) consists of strips of clear polystyrene strung onto a rectangular frame of stainless steel wire, the slides kept from touching one another by means of narrow plastic strips glued across the top of the front face of each. The plastic slides are an advantage when weight is a problem, as in air travel, but this feature also makes their use more complicated, since they must be weighted artificially if they are to make good contact with the materials from which the foraminifera are to be extracted. Discs cut from waste styrofoam sheet and attached to the trays by suitable lengths of string make effective floats to mark the position of the trays when they are placed in deep holding tanks, large aquaria, or directly in the sea itself.

The harvest racks have proved their worth under a variety of conditions. By placing a group of them in the large aquaria and holding tanks encountered around most marine laboratories, one can often procure good numbers of specimens of the several species that generally thrive there. Left on top of sediment concentrates known to be rich in foraminifera, a rich collection of living specimens can be obtained overnight, and successive harvests can be reaped daily for several days or even weeks when adequate precautions against fouling are taken. The same procedure can be used to harvest specimens from smaller laboratory cultures, the racks and slides suitably miniaturized to fit the containers.

Of the three different types of harvest array illustrated here the lowermost (Fig. 8a) is the most generally useful, it being particularly easy to examine in the loaded condition directly under a stereoscopic microscope. In this way a reasonable estimate of a sample's richness can quickly be made. If a more careful examination of individual specimens is desired, the slides can be removed singly from the rack and examined in a shallow plastic tray with built-in ledges that support the slide above the tray bottom and prevent crushing the specimens attached to the slide's under surface. If an extensive harvesting program is anticipated it would be well to plan to work in a room where illumination can be controlled, so that full advantage can be taken of its careful use. The specimens can be seen and the yield generally evaluated with the unaided eye or with the low magnification of a hand lens or a stereoscopic loupe (mounted on the forehead) if light strikes the slides almost parallel to the flat surface and they are viewed against a dark background.

To retrieve specimens that have crawled up the harvest slides, the entire rack should be transferred to a shallow tray where the foraminifera can be brushed from the surfaces of the slides or swept from them with a small rubber squeegee (a piece of windscreen wiper blade with a wire handle). The rack should not be brought into the air during the transfer but kept submerged constantly and handled carefully to avoid dislodging specimens prematurely from their delicate and tenuous pseudopodial moorings. The concentrate ultimately left in the final tray consists of living foraminifera, various invertebrates, diatoms, and bits of debris brought up the slide surfaces by the migrating organisms. If empty tests of foraminifera are abundant in the samples on which the slides were supported, some of these will almost certainly be carried up the slides by the living ones or by other active organisms, but, as a rule, viable specimens far outnumber dead ones or empty tests when this procedure is employed.

Once the specimens have been brushed from the slides into a shallow tray of seawater they can be isolated by mouth pipette and, after suitable washing, transferred directly to the containers in which cultures are to be established. If they are to be introduced into large aquaria already containing a wealth of microscopic metazoans, there is little need to isolate the foraminifera from the minute creatures that climb the slides with them, so the crude concentrate can be placed directly from the slide racks into the tanks; but if more refined cultures are desired, the foreign organisms should be removed at this stage.

V. Culture Techniques and Equipment

A. Types of Culture

The types of culture found to be most practical in the study of living foraminifera are often relatively crude in terms of the more sophisticated protozoological and microbiological standards set today. Practically all of them contain bacteria at least, and many contain other protozoa as well and even such smaller metazoans as nematodes. The culture sources for specimens used in the earlier biological studies on the group appear generally to have harboured other organisms than foraminifera, and even today some students of the group choose to maintain relatively large circulating or non-circulating aquaria in which a few species can thrive, or at least survive, in a balanced community of smaller invertebrates and algae. Generally, the larger the aquarium and the more nearly natural the flow of water through it the greater the variety of species that can be maintained without special attention, but even in non-circulating systems ranging in volume from tens of milliliters to several liters, a few species have proved surprisingly tenacious when accorded only moderate care. As a rule, however, the correlation between the level of productivity achieved in a culture and the amount of energy devoted to its maintenance is gratifyingly high.

The approach used in culturing foraminifera depends, of course, upon the goals of one's culture effort. If this is merely to have large numbers of organisms readily available for other studies—the motivation of most students of the group—then the relatively simple pragmatic and essentially empirical approach still common in protozoological circles during the first third of this century is adequate for several species. If,

however, the problem of cultivation is itself the ultimate consideration rather than the mere means to an end, a more sophisticated
and rigorous program must be developed.

If one wishes merely to maintain large numbers of specimens without
regard to the ancestry of the individual, containers of relatively large
volume (including standard invertebrate aquaria) are preferable to
small ones, but if one needs to maintain genealogical control or repeatedly
examine the same individual throughout its life span, then cultures of
smaller volume should be considered, at least until substantial stocks of
the material can be developed. The care required to maintain small
cultures generally exceeds that for large ones, particularly if the former
are non-circulating, the latter circulating, but the advantages of maintaining individual identity often justify the additional investment.
Another obvious advantage of smaller cultures is that more of them can
be kept in the available space and a greater variety of experimental
conditions can be established and tested simultaneously, thereby increasing the chance of success with a broader range of species. A final advantage is that the loss of a small culture is less drain on one's resources,
including specimens, food organisms, seawater and nutrients.

B. Culture Containers

1. AQUARIA AND LARGE-VOLUME CONTAINERS

The ease with which some species of foraminifera can be maintained in
well-balanced marine aquaria is well known to students of the group;
this approach has recently become popular among paleontologists
(Marszalek and Hay, 1968). Such aquaria are useful as storage containers for stocks, particularly in laboratories remote from the sea, and
the supply of specimens from them, while often hazardous and un-
predictable, can satisfy many needs, especially the simpler pedagogical
ones. The principles governing the selection and maintenance of large-
volume circulating or non-circulating aquaria for the relatively uncritical husbandry of foraminifera are essentially the same as those for
marine invertebrates generally, but the serious student of the subject will
soon find that some basic changes in the design of the container greatly
enhance the efficiency of his culture program. Most commercial aquaria,
for example, are too deep for the easy removal of microscopic benthonic
creatures, to say nothing of the difficulty one encounters in examining the
substratum directly with a microscope to observe and study the organisms *in situ*.

The presence of gravel or sand on the bottom of the circulating
aquarium, though unquestionably important in establishing proper
conditions for both biological and mechanical filtration (see extensive
discussion of the subject by Spotte, 1970), greatly increases the difficulty
of examining, sampling, and harvesting the foraminifera. Because most

foraminifera are themselves considerably smaller than the 2–5 mm. size recommended for the gravel particles of an efficient filtration system, large numbers of them will actually become interstitial and must then be retrieved by removing some of the gravel and applying the usual washing-sieving procedures to it or by placing an artificial substratum on the gravel surface that can be removed and examined after some of the vagile forms have wandered from the interstices onto it. The harvest racks described earlier in this account are one means of accomplishing this. They exploit the habit many species have of climbing any vertical or inclined surface that presents itself. [Schaudinn (1894a) used a horizontal microscope designed by Schulze to examine specimens *in situ* on the walls of an aquarium. Specimens so situated can readily be removed with a pipette for closer examination.] If the productivity of the foraminifera is to be maintained at an adequate level, care should be taken to provide them with the type of food they require and to prevent their destruction by browsing invertebrates or other predators. All in all, then, while large-volume circulating aquaria supporting a well-balanced but relatively complex community of invertebrates and algae can successfully serve to maintain stocks of a few species of foraminifera, the additional time and effort required to maintain numerous small-volume cultures can usually be expected to produce greater, more diverse, and more readily harvested yields.

Containers having a relatively large volume but lacking either circulation or aeration can be surprisingly productive of some of the more hardy laboratory species and are useful in maintaining them for long periods of time (many months or even for years) if: (1) the temperature is kept a few degrees lower than that of the species' natural environment; (2) adequate food is provided; (3) the salinity is kept relatively constant; and (4) the culture community is restricted to protistans and minute nematodes. Under such conditions the water need be changed only once or twice a year; stocks of some species have been maintained in such aquaria for a few years without a change of water.

2. SMALL-VOLUME CONTAINERS

For many types of study the advantages listed earlier for small-volume containers outweigh those for large containers, although where possible both types are desirable. Larger containers are particularly valuable as insurance against the complete loss of all laboratory stocks, but small-volume cultures can be remarkably persistent for some species. *Allogromia laticollaris*, for example, can be kept in a viable condition for a year or more in tightly closed plastic boxes holding only 200 ml. of fluid, and this without a change of the culture medium in which it grows. The present discussion applies essentially to non-axenic culture systems, since the axenic approach has not yet met with success. The problem of obtaining sustained growth and reproduction in the absence of bacteria has even led Lee and his associates to suggest that they are indispensable in the nutrition of these protozoa (Muller and Lee, 1969).

Small-volume, single-species cultures are of two basic types, termed here "mass" and "isolation", the latter containing organisms derived from a single parent (*i.e.*, clonal, in the case of asexually-reproducing or autogamous species) or a single zygote.

Mass cultures, loosely those resulting from the desire merely to produce large stocks without particular regard to, or control of, the ancestry of individuals, are generally best maintained in dishes or boxes that can easily be examined under the stereoscopic microscope, depth, then, being an important consideration. For ease in examining the entire culture and manipulating individual specimens within it, a container no larger than the stage of the microscope is generally easier to handle than a larger one, and round ones are easier to manipulate under the microscope than rectangular ones, although the latter require less storage space. In any case, however, they should stack securely when filled with fluid. Lids of clear polystyrene are preferable to polyethylene ones, because they permit rapid visual inspection without removal; they must fit tightly to prevent evaporation. Ideally, the containers should nest when empty, the slight taper essential to this also being an advantage in examining the sides of the container under the microscope. Clear polystyrene is the material of choice for culture containers of various sizes. The numerous suppliers of suitable boxes can be located through the trade publications of the plastics industry, particularly the annual encyclopaedic issues of *Modern Packaging* and *Modern Plastics* magazines.

Four different types of box with tightly fitting lid are routinely employed in the writer's laboratory: a nesting rectangular one (10 cm. × 10 cm. × 6 cm.) for the long-term maintenance of stock (mass) cultures; a small, non-nesting round one (55 mm. diam., 23 mm. depth) for the initial development of mixed stocks and single-species, clonal, or otherwise-controlled lineages; a round, non-nesting box of intermediate size (83 × 35 mm.) for use when a species has been established and larger numbers are required; and a large, non-nesting, round one (150 × 30 mm.) for special experiments, demonstrations, and the development of still larger stocks. Glass Petri dishes have long been used in various laboratories, but they require more careful handling to avoid spillage, breakage and evaporation; plastic ones have proved particularly vulnerable to evaporation, hence the general preference for boxes with tightly fitting lids.

Freudenthal *et al.* (1963) have described a large system of circulating aquaria for mass cultures, while Arnold (1966) has described a simple apparatus that provides circulation in a closed system of small-volume containers and presented some other approaches to the problem of isolation culture containers. Various workers have used the more conventional microbiological and protozoological methods in developing cultures of extremely small volume; Schaudinn (1894b) and Jepps (1942) have described ways of observing individual specimens for long periods. The need remains for someone to apply intensively the methods and products of plastics technology to the development of culture systems particularly suitable to the requirements of the foraminiferologist.

C. Culture Storage

Cultures of warm-water species are generally the simplest to care for, because they require no refrigeration. They may be placed in a convenient window, preferably with moderate, rather than intense, natural illumination, but a special cabinet wherein temperature and light can be moderated and monitored is a definite advantage. The natural refrigeration of the British Isles and other cooler climes eliminates the need for air conditioning or refrigeration for foraminiferal cultures maintained there, but if the ambient temperature of one's laboratory normally remains much above 20° C., some means of cooling the cultures should be provided for species normally found at comparable or cooler sea temperatures. Permanently installed air conditioning or portable units can be set to provide a temperature suitable to a wide range of species and not unduly uncomfortable for the worker himself (18–21° C.). A cold room (in the 10–15° C. range) is useful for species from colder water, for the long-term storage of stock cultures of cooler-water species, for the temporary storage of freshly collected (raw) samples, and for the long-term storage of stocks of sea water and nutrient media. The simple expedient of blowing air from such a cold room through a series of well-insulated interconnecting cabinets in an adjacent room is a relatively inexpensive way of achieving a broad temperature gradient (ours is 10–23° C.) for large numbers of cultures.

An ordinary domestic refrigerator, modified as earlier described (Arnold, 1954a) or fitted with a thermostatically-controlled tape heater and a fluorescent lamp operated by a timer, is an inexpensive but adequate substitute for the elaborate environmental chambers available commercially.

D. Nutrient and Culture Media

Foraminifera are customarily kept either crudely in large circulating or non-circulating aquaria with other invertebrates or in smaller, more carefully regulated containers of various types. When kept under the cruder conditions, seawater, either natural or artificial, is the usual culture medium. If the system is an open one, most workers assume that the incoming seawater will contain the nutrients essential to the growth of an adequate food supply for the foraminifera, it being generally impractical to enrich a medium destined only for a brief sojourn in the culture container. Generally the encouragement of a moderate growth of diatoms and other unicellular algae makes possible the development of a few species of foraminifera under such conditions.

For most purposes, however, cultures of smaller volume over which more careful control can be exercised are favored. For these, two approaches are immediately apparent: the foraminifera may be inoculated into a culture of food organisms, or the food organisms may be introduced to the containers of foraminifera. In the first case the culture fluid is that of the food culture, but in the second it may be either plain seawater, if further multiplication of the food organisms is not to be encouraged, or a medium suitably enriched, if it is. The choice of procedures depends upon the foraminifer's requirements and the experimenter's needs. Needless to say, the medium used for the food organism must, in any case, not be harmful to the foraminifera. The most commonly employed nutrient medium is Føyn's (1934) "Erdschreiber". This medium, or slight modifications of it, such as those recommended by Lengsfeld (1969) and Heckmann (1963), seems to have been the mainstay of most of the successful foraminiferal culture programs of the past three decades. If the medium gives excessive algal growth when used at full strength, it can readily be diluted to give the desired results; the range of unicellular algae it will support offers a wide selection to the foraminiferologist. If one wishes to operate a simple but effective program of algal cultivation in support of his foraminiferal research, this medium is a dependable keystone for it.

Natural seawater still seems to be the fluid of choice for the preparation of Føyn's "Erdschreiber" and for general use in handling foraminifera, but the increasing interest in their study at inland laboratories has been fostered by the recent development of several synthetic substitutes that are today as readily, though not as inexpensively, available as Gosse's (1856) original one which proved so widely successful for Victorian aquariists generally. Gosse described the use of the early one, Spotte (1970) the more recent ones, particularly in large public aquaria, but the principles apply to small systems as well. In recent years some students of the foraminifera have reported success in the use of synthetic seawater, but the greater ease and economy of using natural seawater will undoubtedly continue to give it favor, particularly in laboratories reasonably close to the coast.

Although the essential principles of effective water management in the cultivation of foraminifera are gradually finding their way into the foraminiferological literature, the approach has long been empirical and pragmatic. The vast body of fundamental knowledge on the subject developed during the past few decades should, however, be carefully examined with the aim of generally increasing the efficiency and effectiveness of foraminiferal culture programs.

E. Food and Feeding

The nutritional requirements of foraminifera vary, not surprisingly, from species to species, but diatoms and other unicellular algae, particularly chlorophyceans, seem to be the most generally useful food organisms. Some species thrive on killed algae. Grell (1954) reported, for example, that *Rotaliella heterocaryotica* does well on heat-killed *Dunaliella*, but another species within the same genus, *R. roscoffensis*, does not. The type of food organism used in maintaining the various species listed in Table 1 is indicated, as is the name of the investigator and the date of his report.

The ideal goal toward which the experimental physiologist strives in developing cultures of foraminifera is to establish axenic conditions, but the more nearly attainable one set by most biologists has been to reduce the food source to a single organism and leave the bacteria—a puristically undesirable but naturally tolerated and artificially (*i.e.*, culturally) tolerable reality of life—to their own devices and fate once a few simple precautions, such as heat-treating or pasteurizing all culture fluids, have been taken. The use of antibiotics to regulate bacterial activity at various stages in the development of cultures of foraminifera and the organisms used as a food source for them has been described by Muller and Lee (1969) and by Lagarde and Arnold (1967). Such strategy is unquestionably valuable for certain specialized culture programs, but the complications introduced by resort to it quickly become formidable, for which reason this approach has not proved popular among students of the group.

Because of the varied needs of the foraminifera it is desirable to have several different species of food organisms on hand at all times. Among these should be *Dunaliella*, *Chlorella*, and one or more species of small naviculoid diatoms. In addition, two or three species of filamentous blue-green algae are useful. Some foraminifera, such as *Allogromia laticollaris*, grow better and can be maintained more easily for long periods of time in diatom cultures containing them than in their absence. I have been unable to achieve sustained growth on them alone, but Lengsfeld (1969) has reported using *Schizothrix calcicola* successfully as the food source for this species.

For many species an effective procedure is to introduce the foraminifera into a culture of the food organism after a relatively uniform turf has developed over which the animals can graze or, in the case of such forms as *Calcituba polymorpha*, upon which they can attach and undergo further growth and development. If maximal yield in minimal time is desired, the method is most effective for vagile forms, the principal danger being excessive algal growth; the culture fluid should be changed relatively often to remove accumulated waste products and to replenish nutrient materials. Harvesting vagile forms from such a uniform turf is a relatively simple matter.

The obvious alternative of introducing the food organism to cultures already containing foraminifera is preferable if for any reason the animals must not be disturbed or if the feeding must be carefully monitored. A third procedure is to introduce food organisms and foraminifera simultaneously and then maintain conditions suitable for the balanced development of both. This method is probably the simplest and fastest way of subculturing stocks or maintaining them over long periods, but the hazards attendant upon it are more numerous than for the other two, as will be discussed under "Subculturing". Nature herself undoubtedly employs all three methods, and species respond differently to them in the sea, just as they do in the laboratory. In any case, the rate of algal growth can be controlled by varying the type and intensity of illumination and the amount of nutrients supplied the cultures.

As cultures are established, a feeding regimen and schedule should be devised that suits the particular culture program. Large cultures will need weekly feeding, whereas smaller ones need be fed only once or twice a month. The amount of food added to the culture should be determined by observing the quantity actually used by the animals between feedings, the critical point being to avoid overfeeding and the consequent accumulation of excess algae in the culture. If living algae are supplied as food the principal hazards are that they overgrow the foraminifera or that their metabolic products create conditions unfavorable for them; if dead (heat-treated) algae are used, the hazard of bacterial multiplication and its consequences must be countered.

F. Other Culture Conditions

1. *Salinity.*—Experience has shown that for most practical purposes the problem of salinity control is secondary to those involving food, temperature, and other organisms, because by changing the culture fluid frequently and either preventing or compensating for evaporation (adding distilled water to a predetermined level in the container) an adequately constant level can easily be maintained. If artificial seawater is employed, particular attention must, of course, be paid the matter of salinity and salt balance, as discussed by Spotte (1970).

2. *Hydrogen-ion concentration.*—The range in pH from 7.5 to 8.3 generally found acceptable for the cultivation of marine organisms is, in practice, seldom exceeded in cultures of foraminifera if proper attention is devoted the other factors of which changes in pH are merely symptomatic rather than causal. In circulating aquaria proper buffering is easily achieved by the use of calcareous gravel in the filter bed, and in closed

systems by the methods of careful husbandry already described, including frequent change of medium and careful control of the food source. A lowering of pH below the acceptable range is more likely to occur in marine systems than is an excessive rise, because bacterial action increases the levels of free carbon dioxide and nitrate, while soluble organic wastes, tending to coat the calcareous gravel of the filter system, probably hamper its buffering action. In circulating systems these hazards are reduced by: (1) suitable biological filtration (nitrification, by autotrophic bacteria; mineralization of organic nitrogenous compounds by heterotrophic bacteria; and denitrification by both types of bacteria); (2) adequate turnover and surface agitation to prevent an excessive buildup of CO_2; (3) proper cleaning of the mechanical and chemical filters; and (4) buffering and the regular replacement of part of the water [Spotte (1970) recommends 10 per cent every two weeks, but see also Clark and Clark (1964).] In closed systems, large or small, a safe pH level can be maintained by regular changes of the fluid and frequent subculturing, a process to be discussed subsequently.

3. *Oxygen.*—In circulating systems with adequate turnover, aeration, and surface agitation, ample oxygen is generally available for foraminifera if larger invertebrates in the system are obviously in good condition. In closed systems and small-volume containers where living algae serve as a food source, oxygen depletion seldom proves problematical, but the procedures in water management for large systems should be adapted for the control of oxygen levels in small ones when the simpler empirical methods presented throughout the present account do not suffice.

4. *Light.*—Little is known of the direct effect light might have on foraminifera, but its obviously critical influence on the algae upon which so many species seem dependent must constantly be borne in mind. Many careful studies of the light requirements of particular algae have been conducted, and sources of information on these are numerous. For the most efficient maintenance of algal stocks as well as the algae in the foraminiferal cultures, attention should be given both the quality and the quantity of light supplied. Foraminiferologists have found that 8 to 16 hours of illumination daily from a 40-watt, daylight-type fluorescent tube placed at a distance of approximately 25 cm. from cultures kept at a temperature of 19–21° C. gives satisfactory results for various species of algae and the species of temperate foraminifera that thrive upon them, but considerable variation around these values is permissible for some of the hardier species, and both natural illumination and light from incandescent bulbs can give satisfactory results when used with caution and discretion.

5. *Temperature.*—Considerable attention has been devoted to this factor by students of the foraminiferal culture environment, because of the

easily observed effects of temperature changes in the laboratory and the obvious correlation between temperature and distributional patterns in nature. The usual procedure is to attempt to duplicate, or at least closely approximate, the temperature at which the organisms live in nature, but seasonal variations, as well as the fact that the intertidal and shallow-water species generally used experience considerable variation in their natural environment, mean a compromise must be effected in selecting the temperature for the laboratory cultures. If the upper end of a species' natural range is selected, more rapid growth and reproduction and, consequently, higher productivity can be expected, but more care is necessary to keep the culture operating efficiently and healthily at the generally accelerated rate. If, by contrast, the lower end of their natural temperature range is selected, then most life processes are commensurately reduced, a condition sometimes preferred, as, for example, when stocks must be maintained with minimal attention for protracted periods. In any case, care must be exercised to determine and distinguish between tolerated and optimal temperatures for the major life processes, vegetative and reproductive alike.

Inasmuch as temperature has an effect not only on the foraminifera but on the food organisms and any other living culture associates, a change in temperature might well be expected to have far-reaching effects on the culture community as a whole. A shift in the dominance from one form to another can easily be caused by a relatively modest change in temperature. The rates of growth and reproduction and, consequently, the generation time are often sensitive to temperature change, and the general balance of the culture community may be drastically upset if this factor is changed appreciably. Since temperature is one of the most easily controlled of factors, many workers choose to keep it constant, thereby controlling at least one of the important variables in the system.

With the technological advances made in heating, refrigeration, insulation, construction materials, and control devices in recent years, one can purchase equipment to satisfy the most fastidious taste, but, fortunately, these advances have also opened new horizons for the ingenious person who prefers to fabricate his own equipment. Ranging from elaborate environmental control chambers with a cost of institutional dimensions to homemade cabinets constructed from scrap and used parts, the possibilities for implementing temperature control are principally limited by one's imagination, budget, abilities, and temperament.

Temperate species of foraminifera and their food organisms are most often kept within the 19–22° C. range, though in warm climates the upper limits may approach 25° C. and for long-term storage the lower may extend to 15° C. The range is, of course, further extended for tropical and cold-water species. If the simple principles of heating and refrigeration are adhered to, the long-term maintenance of cultures within either the broader or the narrower ranges poses few problems that cannot readily be solved at modest cost. By wisely choosing species that thrive at temperatures approximating the average ambient temperature of one's laboratory, the relatively slight complications associated with adequate temperature control add little to one's worries.

6. *Associated organisms.*—Although the enticing and theoretically beguiling goal of axenic cultivation popularized by the microbiologists and many physiological protozoologists has tempted some students of the foraminifera, and a few have devoted considerable energy to the task of achieving it, the more usual and generally attainable goal is the one Grell achieved (1954) in his study of *Rotaliella heterocaryotica*: ". . . gelang es, die Zucht nach und nach rein—wenn auch natürlich nicht bakterienfrei—zu bekommen." This realistic goal of gradually purifying the culture of all living associates other than bacteria (and the food organism, of course) has been achieved by several workers for several species of foraminifera. The simple expedient of heating all culture fluids once or twice to 80–90° C. not only eliminates practically any metazoans and most protozoans but decimates the bacterial population as well, but in the early stages of domesticating new species it is often desirable to set large numbers of cultures rapidly, devoting little time to the removal of protozoa and smaller metazoans in the inoculum from sediment sievings or algal washings. It is at this stage that most of the undesirable forms proliferate excessively. At this stage it is also too early to be certain which of the various species of foraminifera present in the new inoculum will multiply, so it is unwise to devote too much time and attention to the careful evaluation and purification of each culture. The more promising cultures can gradually be selected and the more promising individuals within these gradually removed to heat-treated medium after careful brushing to remove associated organisms. If undesirable associates appear in these cleaner cultures, they can be pipetted or decanted off before their numbers get out of hand, the alternative being to pipette the foraminifera themselves into fresh, clean cultures if the undesirables increase excessively in the intermediate ones. By continuing this process, the promising species and individuals can gradually be isolated from the unwanted residue and an active program instituted for reducing the culture community to its barest essentials within the limits of practicability and need appropriate to one's particular study.

Of metazoans that can safely be left in cultures—at least in the early stages of purification—minute nematodes seem to be among the least troublesome or hazardous, although their frequent occurrence in the tests of dead or dying foraminifera renders even their benignity suspect. Of the protozoa, the taxonomically enigmatical sarcodinian *Trichosphaerium sieboldi* is one of the more frequent and hardy contaminants.

This very hardiness, more than any direct and obviously deleterious effect it might have on the foraminifera, makes it troublesome, because it frequently multiplies more rapidly than the foraminifera and voraciously consumes food essential to them. Moreover, when killings for cytological study are to be as free of foreign organisms as possible, *Trichosphaerium* can be a nuisance. While the bacteria-consuming dinoflagellate *Oxyrrhis marina* probably plays a useful role in the early stages of developing cultures, and some species of foraminifera grow and reproduce effectively in the presence of small numbers of them, this flagellate should certainly be considered a source of potential danger, particularly if it finds its way into the stocks of food organisms or, as Grell (1972) has indicated, into cultures of foraminifera having flagellated gametes. Hauenschild (1972, pp. 230–1) has presented well the case for an association of known organisms rather than axenic or single-species cultures of marine animals generally; the principles he enunciated are equally appropriate for foraminifera.

7. *Substratum.*—For most purposes the clean, smooth surface provided by the glass or plastic of the culture container itself is ideal for the growth and development of foraminifera and the organisms on which they feed. It is also favorable for the low-power examination of the culture with the microscope and conducive to easy maintenance. By means of the simple expedient of placing algal coated coverglasses or glass slides on the floor of the culture containers and letting the foraminifera crawl onto these, the substratum becomes, in effect, detachable, so that it and its epibiota can be examined directly with the high magnification of a monoscopic microscope. The coverglasses or slides can either be placed flat in the dish and later extracted for examination or direct passage into the fluids of cytological preparation, or they can be placed on edge in racks like those already described.

The practice of placing fine sand or other sediment of an inorganic nature on the culture dish to provide a substratum would seem merely to complicate unnecessarily the chore of examining, manipulating and retrieving specimens. In the initial stages of starting cultures from sediment concentrates or the sievings from algal scrubbings, a layer of particulate matter the thickness of one or two sand grains may have been spread over the bottom of many dishes of seawater to permit the foraminifera to migrate to the sediment surface, whence they may be removed by pipette or allowed to migrate onto harvest slides or algal tangles for retrieval; but the separation of the animals from the sediment is essential to efficient subsequent treatment and should not be nullified by the addition of more sediment later. Of course, if an agglutinating species is being cultured and cannot be maintained without an agglutinate, some sediment must be added, but some agglutinating species are well known to be capable of using artificial agglutinates or even to thrive in the complete absence of these, in which latter case they are particularly favorable for microtomy.

G. Subculturing

The crudest cultures of foraminifera, *i.e.*, those used merely as a holding or starting device for freshly collected specimens, may be inoculated merely by spreading a small quantity of sieved sediment or algal scrubbings over the bottom of a dish or aquarium containing seawater and a food source, but as more refined cultures are developed individual foraminifera or concentrations of specimens in varying stages of purification must be inoculated into fresh cultures. The container for the new culture should be well cleaned before use, cleaning materials that might have a deleterious effect on the organisms either being avoided altogether or scrupulously rinsed away after use. If an algal turf has been allowed to develop prior to the addition of the foraminifera, the culture fluid should be replaced at the time of inoculation. If, instead, the food organisms are to be poured or pipetted into the dish containing the foraminifera, the spent culture fluid should be removed and replaced with fresh medium before the food organisms are introduced. The transfer of individual foraminifera is efficiently accomplished through the use of mouth-controlled pipettes, the tip being doubly drawn to give the most precise control of the flow. The principal hazard in using pipettes is that the specimen adheres to the inner wall of the tube or attach itself thereto by its own pseudopodia, but if the water is allowed to rise to its normal level in the tube before the specimens are drawn in and the contents are quickly ejected into the new culture, the chances of loss in this way are greatly reduced.

Prior to inoculation, the foraminifera should be cleaned well by means of a fine sable-hair brush to reduce the risk of introducing undesirable associates from old to new cultures. Swirling the culture container (panning), or swirling in combination with rhythmical tapping of the walls, often yields good concentrations of specimens that are not securely attached to the container, but vigorous brushing is sometimes necessary to loosen those that are. For species that cement themselves even more securely to the substratum, a small chisel-like tool fashioned from plastic or from a stainless-steel dental tool is necessary.

The effect of letting a container of nutrient fluid and food organisms become "conditioned" before the foraminifera are introduced has not been adequately evaluated for all the species that have been maintained in the laboratory, but it is certainly not essential to the development of vigorous populations of several of the more successfully domesticated forms. Spotte's (1970) discussion of the conditioning process in circulating systems, however, contains much background information of use to those who culture foraminifera.

The number of specimens inoculated into a culture depends upon the aims of the culture effort. If isolation cultures, clonal lineages, or other types of cultures with known genealogy are to be instituted, then a single specimen or a few specimens of known parentage suffice, but if large harvests are the sole goal, then a number sufficient to give the desired yield in the time allotted must be introduced, the number varying, of course, with the reproductive rate and brood size of the species, all else being equal. As the population increases, periodic harvesting or subculturing is conducive to large sustained yields, the yield being directly proportional to the amount of energy provided the system (and also, as a rule, to the energy expended on it by the culturist).

Reproduction shortly after the introduction of specimens to a culture is a relatively poor clue to culture conditions, because many species normally reproduce shortly after being brought into the laboratory; their young, however, often fail to grow to reproductive maturity themselves. Reproduction after long periods in culture, however, is generally a good sign.

Cultures designed to produce maximum yields need more attention than those kept as stocks or as insurance against loss of a species or lineage, so subculturing of the former should, naturally, be more frequent. The optimum density for a container depends upon its size and the needs of the particular species. Space sufficient for the full expansion of each individual's pseudopodial network is an obvious requirement for species that tend to remain relatively stationary at maturity, but if its food, in addition to forming a relatively uniform turf over the culture floor, also develops in numerous clumps onto which the foraminifera can climb, a larger population can be sustained than that on a smooth cohesive turf, the critical factors here apparently being grazing space and total quantity of food. The normal successional changes in the living community that result from changes in the physical and chemical properties of the culture environment complicate the task of evaluating the cultures. In the circumstances under which and the purposes for which most workers grow foraminifera it is quite impractical to attempt to monitor and control more than a few of the more obvious factors in the culture environment, but as one's experience increases so will one's ability to judge the point at which a culture should be renewed, subcultured, or replaced. The number and percentage of foraminifera reproducing, actively adding chambers, sending out pseudopodia, and gathering food are all good, practical measures of a culture's productivity and general condition, but many species that exhibit encouraging behaviour during the first few days after inoculation from freshly collected samples go into a rapid decline thereafter and succumb to the still-poorly-understood rigours of the laboratory environment. Until the performance for a species is well known, cultures should be examined daily for any sign of a widespread cessation of pseudopodial activity, feeding, or chamber addition other than that which normally presages reproduction. If these signs are unfavorable, steps should immediately be taken to determine the nature and cause of the problem. Changing or adding culture fluid will often induce renewed activity, if a change in the food source alone seems to be ineffective. The several days immediately following reproduction are usually critical ones for the survival of the newly-produced young. Their activity and general condition at this time are useful clues to the general condition of the culture. If they regularly add chambers at the rate of one every day or so and actively gather food in the interim, all is well, but conditions can deteriorate so rapidly that constant vigilance is essential until the species' performance characteristics can be determined. This is probably not the best time for subculturing, however. It is probably better to wait until most of the animals have reached reproductive maturity or at least adulthood. By exercising some care in the selection of specimens of approximately the same size and age, by feeding and changing the culture fluid often, and by subculturing regularly, a high degree of synchrony can be established in the population's patterns of growth and reproduction, and productivity can be kept at a high level.

If, however, cultures are merely to be kept as a stock from which occasional inocula or study materials are to be taken, a more relaxed regimen can be adopted, which will permit the long-term maintenance of the species with minimal care. By reducing the multiplication rate of the food organism and by keeping the temperature low enough to permit feeding and slow growth but to retard reproduction, stocks of some species can be kept for many months in relatively small containers or for years in larger ones.

H. Harvesting

The procedures for harvesting foraminifera from cultures include those described previously for extracting vagile specimens from sediment concentrates or algal scrubbings and modifications of these that suggest themselves as suitable for the particular culture situations with which one must deal, but they also include those mentioned under subculturing. In some cases the food source in a culture will have been so depleted by the foraminifera that the remains of it can be removed by pipetting or by judicious swirling and pipetting. An aspirator pipette operated by vacuum and equipped with suitable traps and filters can increase one's efficiency if large volumes must be handled.

I. Observation

The effective examination, evaluation and maintenance of a culture of foraminifera requires the competent use of both the stereoscopic and the monoscopic microscope. Ideally one should be able to examine the cultures directly with the microscope rather than by removing elements of it to the microscope, hence the author's general preference for containers that permit this. The simplest solution is to use containers that can be placed on the stage of a stereoscopic microscope and examined with objectives ranging from $1 \times$ to $12 \times$, but if larger containers must be used, other means must be devised. An inverted microscope can be a valuable aid, as can a horizontal microscope and a stereoscopic microscope mounted on an overhead rail along which it can be moved to cover an array of trays.

In any case, careful consideration should be given the selection and use of the microscope and the illuminator it requires. Ideally two different types of illuminator should be employed, both of high intensity and regulated by rheostat. The one for use with the stereoscopic microscope should illuminate a large field (the type with focusable parabolic reflector is exceptionally good). When examining the culture directly, the angle of incidence should be carefully adjusted to take full advantage of the contrast while avoiding unnecessarily high intensity and the heat it generates.

VI. References

Adshead, P. C. (1967). *Micropaleontology* **13**, 32–40.

Angell, R. W. (1967). *J. Protozool.* **14**, 566–574.

Arnold, Z. M. (1954a). *J. Paleont.* **28**, 404–416.

Arnold, Z. M. (1954b). *Contr. Cushman Fdn foramin. Res.* **5**, 1–13.

Arnold, Z. M. (1955). *Univ. Calif. Publs Zool.* **61**, 167–252.

Arnold, Z. M. (1962). *Micropaleontology* **8**, 515–518.

Arnold, Z. M. (1964). *Univ. Calif. Publs Zool.* **72**, 1–78.

Arnold, Z. M. (1965). *In* "Handbook of Paleontological Techniques" (B. Kummel and D. Raup, eds.) pp. 420–422. Freeman, San Francisco.

Arnold, Z. M. (1966). *Micropaleontology* **12**, 109–118.

Arnold, Z. M. (1967). *Arch. Protistenk.* **110**, 280–304.

Bradshaw, J. S. (1955). *Micropaleontology* **1**, 351–358.

Buchanan, J. B. and Hedley, R. H. (1960). *J. mar. biol. Ass. U.K.* **39**, 549–560.

Chinn, A. F. (1972). Unpubl. M.S. thesis, Geol. Dept., Dalhousie Univ.

*Clark, J. R. and Clark, R. L. (1964). *Res. Rep. U.S. Fish. Wildl.* Serv. **63**, 192 pp.

Dujardin, F. (1835). *Ann. Sci. Nat.* (Ser. 2) **4**, 343–377.

Føyn, B. (1934). *Arch. Protistenk.* **83**, 1–56.

Føyn, B. (1936a). *Bergens Mus. Årb.* **2**, 1–22.

Føyn, B. (1936b). *Arch. Protistenk.* **87**, 272–295.

Freudenthal, H. D., Lee, J. J. and Pierce, S. (1963). *Micropaleontology* **9**, 443–448.

Gervais, P. (1847). *C. r. hebd. Séanc. Acad. Sci., Paris* **25**, 467–468.

*Gosse, P. (1856). "The Aquarium", 2nd edit. van Voorst, London.

Gougé, J. (1971). M.A. Thesis, Paleontology Dept., Univ. of California, Berkeley.

Grell, K. G. (1954). *Arch. Protistenk.* **100,** 268–286.

Grell, K. G. (1957). *Arch. Protistenk.* **102,** 147–164.

Grell, K. G. (1958). *Arch. Protistenk.* **102,** 449–472.

Grell, K. G. (1959). *Arch. Protistenk.* **104,** 221–235.

*Grell, K. G. (1972). *In* "Research Methods in Marine Biology" (C. Schlieper, ed.), pp. 248–254. Univ. Wash., Seattle.

Hauenschild, C. (1972). *In* "Research Methods in Marine Biology" (C. Schlieper, ed.) pp. 216–235. Univ. Wash., Seattle.

Hedley, R. H. (1958). *Proc. zool. Soc. Lond.* **130,** 569–576.

Hedley, R. H. (1962). *N.Z. Jl Sci.* **5,** 275–289.

Hedley, R. H., Ogden, C. G. and Wakefield, J. St. J. (1973). *Bull. Br. Mus. nat. Hist.* (Zool.) **24,** 467–474.

Hedley, R. H., Parry, D. M., and Wakefield, J. St. J. (1967). *Jl R. microsc. Soc.* **87,** 445–456.

Hedley, R. H., Parry, D. M., and Wakefield, J. St. J. (1968). *J. nat. Hist.* **2,** 147–151.

Hedley, R. H. and Wakefield, J. St. J. (1967). *J. mar. biol. Ass. U.K.* **47,** 121–128.

Heckmann, K. (1963). *Arch. Protistenk.* **106,** 393–421.

Jepps, M. W. (1942). *J. mar. biol. Ass. U.K.* **25,** 607–666.

Kleinpell, R. M. (1971). *J. of the West* **10,** 72–101.

Lagarde, E. and Arnold, Z. M. (1967). *Vie Milieu* **1(A),** 27–45.

Le Calvez, J. (1938). *Archs Zool. exp. gén* **80.** 163–333.

Le Calvez, Y. (1972). *Ann. Paleont.* **58,** 129–134.

Lee, J. J., McEnery, M., Pierce, S., Freudenthal, H. D. and Muller, W. A. (1966). *J. Protozool.* **13,** 659–670.

Lee, J. J. and Pierce, S. (1963). *J. Protozool.* **10,** 404–411.

Lengsfeld, A. M. (1969). *Helgoländer wiss. Meeresunters.* **19,** 230–261.

Lloyd, W. A. (1858). "A list, with descriptions, illustrations, and prices of whatever relates to aquaria". 162 pp. London.

Lloyd, W. A. (1875). "Official Handbook to the Marine Aquarium of the Crystal Palace Aquarium Company (Ltd.)". 6th ed. The Crystal Palace Company, London.

Marszalek, D. S. (1969). *J. Protozool.* **16,** 599–611.

Marszalek, D. S. and Hay, W. W. (1968). *J. geol. Educ.* **16,** 159–163.

Muller, W. A. and Lee, J. J. (1969). *J. Protozool.* **16,** 471–478.

Murray, J. W. (1963). *J. mar. biol. Ass. U.K.* **43,** 621–642.

Myers, E. H. (1933). *Science* **77,** 609.

Myers, E. H. (1936). *Jl R. microsc. Soc.* **56,** 120–146.

*Myers, E. H. (1937). *In* "Culture Methods for Invertebrate Animals" (P. S. Galtsoff, ed.) pp. 93–96. Comstock, Ithaca.

*Myers, E. H. (1945). *In* "Rept. of the Committee on Marine Ecology as Related to Paleontology" **5** (1944–5), pp. 24–28. Nat. Res. Council, Washington, D.C.

Nyholm, K-G. (1955). *Zool. Bidr. Upps.* **30,** 475–484.

Nyholm, K-G. (1956). *Zool. Bidr. Upps.* **31,** 483–495.

Nyholm, K-G. (1961). *Zool. Bidr. Upps.* **33,** 157–196.

Ross, C. A. (1972). *J. Protozool* **19,** 181–192.

Röttger, R. (1972). *Mar. Biol.* **17**, 228–242.

Schaudinn, F. (1894a). *Sber. Ges. naturf. Freunde, Berl.* 14–22.

Schaudinn, F. (1894b). *Z. wiss. Mikrosk.* **11**, 326–329.

Schultze, M. S. (1854). "Über den Organismus der Polythalamien (Fora-
miniferen) nebst Bemerkungen über die Rhizopoden im allgemeinen."
Engelmann, Leipzig.

Schultze, M. S. (1856). *Arch. Anat. Physiol.*, 165–173.

Schultze, M. S. (1860). *Arch. Naturgesch.* **26**, 287–310.

Schultze, F. E. (1875). *Arch. mikrosk. Anat. Entw. Mech.* **11**, 94–139.

Schwab, D. (1970). *Z. Zellforsch.* **108**, 35–45.

Sliter, W. V. (1970). *Contr. Cushman Fdn. foramin. Res.* **21**, 87–99.

*Spotte, S. H. (1970). "Fish and Invertebrate Culture." Wiley-Interscience,
New York.

Stephenson, W. and Rees, M. (1965). *Pap. Dep. Zool. Univ. Qd.* (10) 205–224
and (12) 239–255.

Stinemeyer, E. H. (1965). *In* "Handbook of Paleontological Techniques"
(B. Kummel and D. Raup, eds.) pp. 276–283. Freeman, San Francisco.

Strong, F. E. (1969). *Turtox News* **47**, 274–275.

Swanson, R. S. (1965). "Plastics Technology." McKnight and McKnight,
Bloomington.

Walton, W. R. (1952). *Contr. Cushman Lab. foramin. Res.* **3**, 56–60.

Weber, H. (1965). *Arch. Protistenk.* **108**, 217–270.

Yonge, C. M. (1949). "The Sea Shore." Collins, London.

An asterisk (*) indicates those references that are of greatest general
interest or use in collecting and culturing foraminifera.

Towards Understanding the Niche of Foraminifera*

JOHN J. LEE

*Department of Biology, City College of City University of New York,
Convent Avenue at 138th Street, New York, N. Y. 10031*

I. Introduction 208
II. Field Approaches 210
  A. Dynamic Aspects 210
  B. Patchiness, Substrate Preferences and Seasonality . 212
  C. Community Related Studies 216
III. Laboratory Studies 219
  A. Culture Methods 219
    1. Agnotobiotic Cultures 219
    2. Gnotobiotic Cultures 221
      a. Fortuitous Methods 221
      b. Inductive Methods 222
      c. Reductive Methods 224
  B. Niche Related Studies 224
    1. Abiotic Studies 224
      a. Temperature, Salinity, pH 224
      b. Multivariate 225
      c. Effect on Food 229
      d. Dynamic Aspects 230
    2. Biotic Studies 231
      a. Food 231
        i. Growing Foraminifera with Food . . 231
        ii. Heat Killed Food 232
        iii. Tracer Feeding Studies . . . 233
        iv. Gnotobiotic Growth Experiments . 236
      b. Density Dependent Interactions . . 238
        i. Intraspecific 238
        ii. Interspecific 239
      c. Niche Representation 241
      d. Ecological Efficiency 244

* Much of the research upon which this review is based was aided by grants from the U.S. N.S.F. and contracts from the U.S. A.E.C. (Reference no. COO 3254–15).

  IV. Symbiosis in Foraminifera 245
        A. Symbionts in Planktonic Foraminifera   .   .   . 245
        B. Symbionts in Benthic Foraminifera 246
        C. Biology of *Chlamydomonas hedleyi* 248
        D. Speculation on the Origin of Symbiosis in Foraminifera. 248
        E. Some Current Problems 251
   V. Concluding Remarks 252
   VI. Acknowledgments 255
  VII. Glossary 255
 VIII. References 257

"I remember when I was at Lilliput, the Complexions of those diminutive People appeared to me the fairest in the World: And talking upon this Subject with a Person of Learning there, who was an intimate Friend of mine; he said, that my Face appeared much fairer and smoother when he looked on me from the Ground, than it did upon a nearer View when I took him up in my Hand, and brought him close; which he confessed was at first a very shocking Sight. He said, he could discover great Holes in my Skin; that the Stumps of my Beard were ten Times stronger than the Bristles of a Boar; and my Complexion made up of several Colours altogether disagreeable: . . ."

—Swift, Jonathan. *Gulliver's Travels; A Voyage to Brobdingnag*

I Introduction

Perspective is one of many illusionary aspects of life. Micropaleontologists, paleoecologists, physical, geological, and biological oceanographers, protozoologists, and marine microbiologists very often look at foraminifera from quite different vantage points. Decades of field and laboratory study suggest that environmental factors such as current, salinity, sediment quality, and, most importantly, temperature, control the overall distribution patterns of individual species or assemblages. Within recent years methods of sampling and analyzing data have been refined so that we have a new perspective on foraminiferal distribution. We now know that their distribution is very patchy within broad ecological zones defined by earlier workers; hundreds of thousands of littoral benthic foraminifera may be found in a bloom at one spot in a marsh and <0.5 m away few, if any, can be found. Even studies with samples taken every 5 m are grossly too far apart when one is interested in the niches of individual foraminiferal species. As a microbiologist, I am awestruck by the inextricable complexities of even 1 cm^3 in the salt marsh epiphytic or epibenthic communities we have been studying. Within 1 cm^3 live perhaps 10^5–10^6 organisms belonging to several hundred different species (Lee *et al.*, 1973; Table I). Within this "wondrous strange" world lie the secrets of the foraminiferal niche. One

TABLE 1

Number of organisms found in an idealized cm³ sample of
a community epiphytic on *Enteromorpha* in an unpolluted
salt marsh

Organisms	Concentration	Number of species in one sample	Total number of species in marsh
Diatoms	1×10^3–10^4	75–100	$\sim$250
Chlorophytes	10–10^3	5–10	>30
Enteromorpha zoospores	1×10^3	varies	
Dinoflagellates	10–100	2	$\sim$30
Prasinophytes	1–2	>1	
Euglenophytes	10–10^2	varies	$\sim$6
Blue greens	10–10^3	2	
Yeasts	5–10	>1	>10
Bacteria	$\sim$$10^4$–$10^5$	$\sim$20	$\sim$75
Zooflagellates, colorless dinoflagellates, Euglenoids	10–10^2	$\sim$4	>50
Amoeba	$\sim$$10$–$10^2$	2	>12
Heliozoa	rare	1–2	$\sim$2
Foraminifera	0–10^3	2–15	>40
Ciliates	10–10^2	4–8	>50
Gastrotrichs	<10	1	
Rotifers	<10	2	
Nematodes	10–10^2	3–5	>40
Microcrustacea	0–10	varies	
Gastropods	$\sim$1	1	
Other micrometazoa	<10		

wonders: why are there so many different species of bacteria, algae, foraminifera and other protozoa, and micrometazoa in salt marsh communities? What are the biological bases underlying this huge diversity of species? What special adaptations do foraminifera have to live in the various zones in the marsh? How large is, and what constitutes, the "universe" for individual foraminifera? What are the key factors which favor the growth of one foraminiferal species over that of another? What role do foraminifera play in the carbon budget and nutrient cycling of the detritus based communities in which they live? To what extent do the niches of foraminifera overlap those of other microherbivores?

If we had the answers to many of these questions we would know a lot more about the biology of the foraminifera and the microbial communities in which they live than we do today. Even with the most up-to-date methods and tools (Rosswall, 1973) the task of answering these questions seems staggering.

The concept of niche is not easily understood except in very abstract terms. Often micropaleontologists and protozoologists think of niche in a spatial or habitat sense and commonly only from data gathered at a single point in time. While these studies are valuable in themselves, they give only very little insight into the biotic constraints which are important in the dynamic processes which establish the realized niches of foraminifera (see Vandermeer, 1972, for an up-to-date review of niche theory). It is my hope in this review to give a microbiologist's microscale perspective of the functional roles of a few littoral benthic foraminifera in their communities and the methods which can be used to gain insight into this aspect of their biology.

II Field Approaches

A. Dynamic Aspects

Field studies are the point of departure for any investigation of the foraminiferal niche. A number of studies in recent years have been concerned with operational definitions of habitat which incorporate a sense of dynamic change and conceptualization of the operational habitats of foraminifera. In salt marshes littoral foraminifera are exposed to continuous changes in air and water temperature, light intensity, DO, pH, Eh and salinity. Bradshaw found significant hourly, daily and seasonal differences in the abiotic variables he continuously monitored in Mission Bay, San Diego, California from January 1964 to the end of August the following year (Bradshaw, 1968). As was expected, tidal and diurnal rhythms were responsible for most of the variations in abiotic factors. Flood tide covered this marsh with sea water from the bay which had a salinity of $\sim$34‰ and a pH of 8.0–8.2. At ebb tide the salinity of the water increased up to $\sim$50‰ and pH went as low as 6.8 (night) or as high as 8.5 (day). Sunrise resulted in an immediate rise in air temperature from 9.5° C to 14° C. Shortly afterward oxygen and pH values increased. The amount of rise depended, of course, on cloud conditions and season (typical light intensity $\sim$110000 lux). Maximum air temperature occurred in the early afternoon (1300–1430 h) while water temperature reached maximum 1 or 2 h later. In the morning there was a positive correlation between light intensity, pH, and oxygen. In the afternoon the values of the latter 2 variables continuously declined until before dawn the following morning when minimum values were reached. At all stations Bradshaw found significant differences between the pH and Eh of the overlying water, the sediment surface, and various depths in the sediment. Although he found only slight differences in the salinity and temperature of stations located approximately horizontally from each other, he did find differences in pH and oxygen values. The highest pH (9.3) was measured among filaments of *Enteromorpha* in the full sunlight, whereas 10 cm away in the shade the pH was only 8.5. Over the year the air temperature varied from 0.5° C in December to 36° C in

October (Fig. 1). Water temperature varied from 4.9° C to 33° C (Fig. 2). His salinity recording apparatus was inoperable much of the time but during the 2 months in which continuous recordings were available the upper (47 and 50‰), lower (30 and 31.5‰), and median (37 and 39‰)

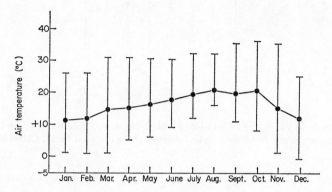

Fig. 1. Changes in air temperature in Mission Bay Marsh (California) measured continuously for a year. Values given are monthly mean (circles) and range (horizontal bars). (After Bradshaw, 1961).

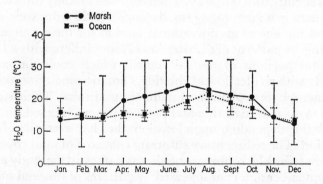

Fig. 2. Changes in water temperature in the adjacent ocean and Mission Bay Marsh (California) measured continuously for a year. Values given are month mean (circles and squares) and range (horizontal bars). (After Bradshaw, 1961).

salinities varied ∼2‰. He did not find seasonal correlations in the changes of pH, Eh or dissolved oxygen (Fig. 3).

On a much larger scale Lynts (1966) in a similar approach studied shallow benthic samples from 19 stations in Buttonwood Sound, Florida Bay. The stations were occupied at 3 day intervals in August and again in February. He also found daily, seasonal and horizontal differences in temperature, salinity, pH, and Eh at the same station.

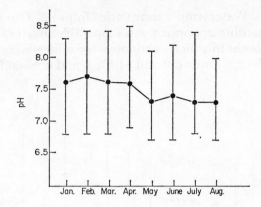

Fig. 3. Changes in pH in the Mission Bay Marsh measured continuously for 8 months. Values given are monthly mean (circles) and range (horizontal bars). (After Bradshaw, 1961).

B. Patchiness, Substrate Preferences and Seasonality

Lee *et al.* (1969) and Matera and Lee (1972) have approached the same problem, that of environmental grain (see review by Vandermeer, 1972), in a slightly different way. Their aim was to study the distribution of foraminifera in a great many small samples (0.28 g dry weight) which approached the size of an operational habitat for these animals. They were looking for patterns of distribution of foraminifera within relatively small littoral portions of the salt marsh which could be correlated horizontally with rivulet flow at ebb tide, climatic conditions, tidal cycle, specific macrophyte growth, and sediment grain size. Their study area was ~1.5 m deep at high tide and parts of it were exposed for ~30–50 min at ebb tide depending upon location, the characteristics of the tide, and the wind. The sedimentary substrate consisted of sand covered with a fine layer of silt. Macrophyte growth was patchy in the study area from May to September and characterized by patterns of seasonal succession. If the spring was mild *Enteromorpha* grew and covered the study area by early May. By July and early August *Polysiphonia* sp and *Zanichellia* became increasingly abundant. In late summer *Ulva lactuca*, *Cladophora* spp and *Ceramium* spp became more prominent at this field station. Some *Fucus* and *Codium* were also found. Storms in late August or September usually scoured the area and removed the macrophytes which were renewed the following spring and summer. In the deep creeks adjacent to the field station *Zostera* was abundant.

They found that the temperatures at the surface generally were identical throughout their study site at high tide but at low tide thermal gradients were established. Depending upon the time of day, tidal cycle, wind, cloud cover, etc. a 5–7° C thermal gradient was established between emergent and submergent (∼1 cm) regions. Coldest water was found in the outflowing rivulets. Salinity gradients of 2–3‰ were also correlated with emergence and current flow as were sedimentary patterns. The finest sediments were located where there was minimum current at ebb tide.

Foraminifera were very unevenly distributed in the small sized samples (∼1 cm³; ∼0.28 g dry weight) taken from the environment. In each of the 3 years studied a very small number of samples contained more than half of the total foraminifera harvested. More specifically: in 1966, 9 of 86 samples contained 61% of the foraminifera; in 1967, 17 of 142 samples contained 81% of the foraminifera; and in 1968, 14 of 530 samples contained 56% of the foraminifera harvested. The population structure of the foraminiferal assemblages was, however, quite distinctive in each of the years studied. They inferred that climatic conditions underlie the differences; 1966 was the last year of a prolonged drought, 1967 a very cold and rainy summer, 1968 a "typical" summer.

In both 1966 and 1967 *Ammonia beccarii* was twice as abundant as any other species collected and comprised approximately 40% of the total foraminiferal population in both years. It was the dominant form in 18 of the 26 blooms (>50 foraminifera per small sample) collected in both years; *Protelphidium tisburyensis*, *Allogromia laticollaris*, *Elphidium incertum*, *Quinqueloculina seminulum*, and *Trochammina inflata* being the dominant forms in the rest. Not unexpectedly, because the marsh had been subjected to drought for the previous 4 years, the species diversity of the foraminiferal assemblage of the community was quite low. Although usually very rare in field collection, *Allogromia laticollaris* was the dominant form in 3 of the 1966 blooms. *Elphidium subarticum* and *Rosalina leei* were common in 1966, rare in 1967, and not harvested in 1968. In the 1968 collections, which were not quite comparable to the earlier 2 years because they included both epiphytic and sediment samples, *Elphidium incertum* (46.6%) and *Protelphidium tisburyensis* (25.6%) were the dominant species. Some species, *Ammobaculites dilatatus*, *Elphidium advenum*, *E. galvestonense*, *E. gunteri*, *Quinqueloculina lata* and *Jadammina macrescens* were found in the epiphytic community but not below in the sediment. *Protelphidium tisburyensis* was a dominant species in the epiphytic community and fairly common as an epibenthic form. *Trochammina inflata*, *Quinqueloculina seminulum* and *Ammotium salsum* were much more abundant in the sediments than in the water column above. *Elphidium incertum* was abundant in both fine grain sediments (median grain size distribution 0.1 mm) and in the epiphytic communities above them. *T. inflata* distribution clustered around a median grain size of 0.46 mm. In the psammolittoral communities studied, the distribution of the latter 2 species was correlated with vertical and horizontal changes in grain size.

Matera and Lee (1972) found no evidence to indicate that foraminifera migrate through the sediments as a function of the tidal cycle even though they had demonstrated earlier (Lee *et al.*, 1969) that these animals have the ability to do so. Taken as a whole, and considering the fact that only 60 of 530 samples studied were cores, it appears that foraminifera are more abundant in the psammolittoral community than they are in the epiphytic.

As discussed earlier, the epiphytic community at Towd Point is quite seasonal (~May–September) and characterized by rapid changes in the macrophyte standing crop. Seasonal changes have also been found in all of the other biotic assemblages which have been studied thus far. In the epiphytic community at Towd Point, *Protelphidium tisburyensis* has most often bloomed in early June, followed by *Ammonia beccarii* in July, and by *Ammotium salsum, Elphidium incertum, E. translucens, E. clavatum* and *Quinqueloculina* spp from mid-July to mid-August (Fig. 4). In the psam-

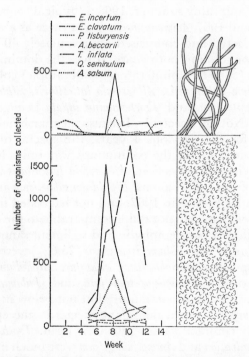

Fig. 4. Changes in the standing crop of foraminifera in the epiphytic community growing on *Enteromorpha* and other macrophytes (top part of graph) and in the psammo-littoral community directly below (bottom part of graph) during the summer at station 19, Towd Point Marsh, North Sea Harbor, Southampton, N. Y. Time is measured in weeks from June 13, 1968, when the epiphytic community first developed at the station. Characteristic storms scoured the area and destroyed the community after the September 3 collection. (After Matera and Lee, 1972).

molittoral community *Elphidium incertum, Ammotium salsum* and *Trochammina inflata* followed each other (Lee *et al.*, 1969, Matera and Lee, 1972). Data obtained by many others lead one to believe that the Towd Point foraminiferal assemblage is a fairly typical one. Buzas (1968), using an elaborate multiple sampler which takes 36 simultaneous contiguous samples, has shown that some of the same species are also patchy in their distribution in Rehobeth Bay. In a succeeding study he (Buzas, 1969) studied spatial homogeneity of 4 species of foraminifera in the same bay using a more widely spaced grid system consisting of 16 stations, each 10 m apart in a 4 × 4 grid. Five replicate samples were taken at each station. Analyses of differences in the populations found at each station led to the conclusion that *A. beccarii, P. tisburyensis,* and *E. clavatum* were very patchy in their distribution, but except for the latter organism patch size was not easily defined. Others working at the same relative sampling scale (Ellison and Nichols, 1970, Lutze, 1968, Lynts, 1966, and Shifflet, 1961) also reported spatial patchiness in their collections.

Seasonal changes in the productivity of foraminifera and in the standing crop of foraminifera are perhaps also very common if we can judge from the reports of Boltovskoy (1964), Boltovskoy and Lena (1969), Brooks (1967), Buzas (1965), Jepps (1942, 1956), Murray (1963), Myers (1943) and Phleger and Lankford (1957) among many others. Though we have yet to evaluate the impact of seasonal changes in the associated floral, microfloral and faunal assemblages of the community on the distribution of foraminifera some elements of the biotic aspects of the puzzle are being assembled on the Towd Point community (Lee *et al.*, 1969, 1973; Lee *et al.* ms submitted for publication, and Kennedy, work in progress). Large standing crops of foraminifera were found in the sublittoral epiphytic communities of *Enteromorpha intestinalis, Zostera marina, Zanichellia palustris, Ulva lactuca, Polysiphonia* spp and *Ceramium* sp as they occurred seasonally but only rarely in the epiphytic communities associated with *Fucus* or *Codium* (Lee *et al.*, 1969). Statistical treatment of the data on the distribution of the most abundant species gave some indication that *Protelphidium tisburyensis* was found most frequently on *Enteromorpha* whereas species of *Quinqueloculina* were less likely to be so distributed. *Ammonia beccarii* and *Elphidium* spp showed little substrate preference. Patches of decaying *Enteromorpha* had the greatest standing crop of foraminifera and a relatively low species diversity index (0.581) when compared to young green patches (0.94). Indicies for *Zostera, Zanichellia, Polysiphonia, Fucus, Ulva* and *Codium* were 0.82, 0.99, 0.86, 0.70, 0.77 and 0.196 respectively.

Under drought conditions much more epiphytic community grew on the macrophytes than during "rainy" or "normal" years (measured as g dry weight epiphytes per g dry weight host macrophyte). The ratio of standing crop of epiphytes to macrophyte was lower at some field stations; possibly explained by stronger currents. There was no correlation between epiphytic community biomass and foraminiferal standing crop. Probability and coincidence tables were constructed with the aim of testing whether various species of foraminifera were distributed at random to each other or whether some shared common distribution patterns. Those which were distributed together have some common "microenvironmental dependence". Thus *Ammonia beccarii* grew well in the communities also favorable to *Elphidium incertum, E. translucens, E. clavatum, E. subarticum, Quinqueloculina seminulum, Miliammina fusca, Rosalina leei* and *Protelphidium tisburyensis*, whereas there were negative correlations between *R. leei* and *E. translucens*; *E. subarticum* and *P. tisburyensis*; and *Trochammina inflata* and *Q. seminulum. Allogromia laticollaris* gew independently of any other foram species.

C. Community Related Studies

The epiphytic diatom assemblage at Towd Point is as complex as any comparable one which has been studied (see reviews by Round, 1971, Aleem, 1950 and McIntire *et al.*, 1971). During the summer months it is an important contributor to the primary productivity of the community (Lee *et al.*, 1973). In one summer (1971) more than 218 species or varieties of diatoms were recognized growing on *Enteromorpha* at field station 19 (Lee *et al.* ms. *op. cit.*), a station which consistently harbors large numbers of epiphytic and psammolittoral foraminifera (Lee *et al.*, 1969a, Matera and Lee, 1972). Six species dominated the epiphytic community during the summer months, accounting for nearly 40% of the total diatom community. Together these species, *Fragilaria construens, Cocconeis scutellum, C. placentula, Achnanthes hauckiana* varieties, *A. pinnata* and *Amphora coffeaeformis* (var. *acutiuscula*), and 6 others, *Melosira nummuloides, Ophphora martyi, Synedra fasiculata* var *tabulata, S. affinis, Navicula platyventris* and *N. pavillardi* comprised more than half the epiphytic diatom population. None of these species was uniformly distributed in the assemblage throughout the summer (Fig. 5). Some species, e.g., *M. nummuloides, S. affinis, Navicula tripunctata* var. *schizonemoides, Nitzschia sigma, N. hungarica* and *N. panduriformis* were more abundant in the early summer weeks while others, e.g., *Cocconeis scutellum, Achnanthes hauckiana* (var. C and D), *A. pinnata, A. chilensis*, and *A. pseudosolea* bloomed in late summer (Fig. 6). Other species, e.g., *F. construens, C. placentula, O. martyi, A. hauckiana, Nitzschia subtilis* (var. *paleacea*) and *Nitzschia amphibia* had several midsummer peaks of abundance. Many species showed high correlation in their abundance or in their appearance in the community, indicating possible abiotic and/or biotic dependencies.

Field studies also indicate seasonal changes in the nutrients available and in the abilities of the epiphytic assemblages to use them (Lee *et al.*, 1973 and ms *op. cit.*). Chemical analyses of the nutrients within and without *Enteromorpha* communities show significantly higher levels of dissolved nitrate (3.3 vs 2 μg atom/1), SiO_2 (42 vs 14 μg atom/1), and

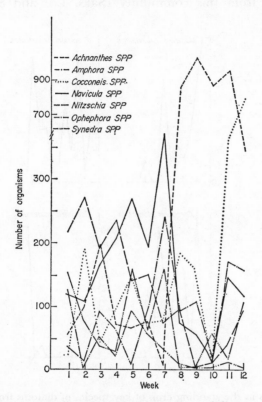

Fig. 5. Changes in the standing crop of major genera of diatoms in the epiphytic communities of *Enteromorpha* at the same station as the foraminifera in Fig. 4. (Plotted from data in Lee, McEnery, Kennedy and Rubin, ms submitted for publication). Plot represents > 1000 diatoms counted in an aliquot of a foraminiferan bloom used as an inoculum for a study of diatom nutritional patterns.

total organic nitrogen (60 vs 32 μg atom/1) within the sea water in the community (Lee *et al.* ms *op. cit.*). As might be expected, the levels of these nutrients fluctuated considerably over the period measured. Experiments with radionuclide labeled substrates (Lee *et al.* ms *op. cit.*, and Kennedy, work in progress) have shown that the changes in the population structure of the community were also reflected by changes in the nutritional pattern, particularly as judged by the qualitative and quantitative differences in the growth of diatoms and bacteria on the differential media tested. The auxotrophic profile of the diatom assemblage suggests that some measure of stability within this important component of the community is achieved through heterotrophic diversity. Laboratory studies on the nutrition (e.g., Fig. 7), carbon budgets, and release of organic materials from 11 selected species of diatoms and chlorophytes from this community (Saks, Lee and Stone, work in

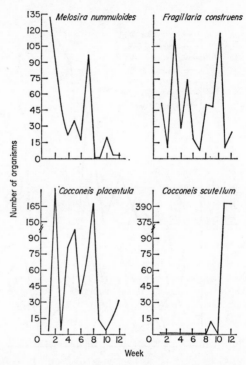

Fig. 6. Changes in the standing crop of key species of diatoms from the same community as in Figs. 4 and 5. (Plotted from data in Lee, McEnery, Kennedy and Rubin, ms submitted for publication).

progress) and the studies of many others (Bunt, 1969, Burkholder, 1963, Carlucci and Bowes, 1970a, b, Fogg, 1966, Hellebust, 1965, 1970, 1971, Hellebust and Guillard, 1967, Hellebust and Lewin, 1972, Huntsman, 1972, Johannes and Webb, 1965, Lewin and Hellebust, 1970 and Lewin and Lewin, 1960, 1967), suggest that many of the organic substrates used by individual species originate within and are recycled over very short distances among the community members.

Though it would seem to the uncritical observer that there has been no dearth of papers on patchiness, seasonality, and "microclimatological factors" and the distribution of various species of near shore foraminifera,

we really know very little about the realized niches of any group of foraminifera from any locality in the world. Hopefully in the future we will
have enough field and laboratory data to describe the realized niches of
a few foraminiferal species. It staggers the imagination to consider just
the field work involved: detailed microclimatological data following the
approaches of Bradshaw (1968) and others; coordinated with extensive

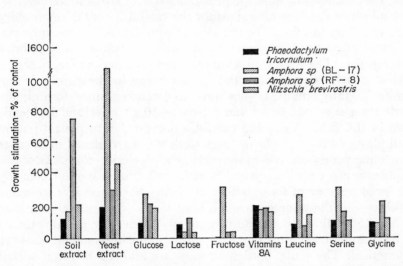

Fig. 7. Stimulation of the growth of 4 representative clones of epiphytic diatoms isolated
from the Towd Point salt marsh. Growth stimulation is measured as enhancement
of growth (100%) over that obtained in an unenriched artificial sea water medium
comparable to unenriched sea water from Towd Point. (Data from Saks, Lee and
Stone, ms in preparation).

patch size and common microenvironmental data along the lines of
Buzas (1965), Lynts (1966), Lee *et al.* (1969), Matera and Lee (1972);
and coordinated with detailed assessments of the distribution and population dynamics of the associated microbial, meiofloral, and meiofaunal
assemblages of the community (Lee *et al.* ms *op. cit.*, Fenchel, 1968, 1972,
Fenchel and Jannson, 1966, Hargrave, 1970, Tietjen, 1969, Teitjen
et al., 1970, Tietjen and Lee, 1972, etc.). The bulk of the field work thus
lies ahead.

III. Laboratory Studies

A. Culture Methods

1. AGNOTOBIOTIC CULTURES

Laboratory studies invariably begin with the culture of the organism
one is interested in. The aims of the study should be carefully defined
before a sampling and culture program is initiated. Agnotobiotic

cultures (cultures in which the identities of the associated organisms are unknown) are the easiest to set up but they are notoriously undependable and often less productive. For many purposes such as life-cycle or some types of physiological studies in which the effects of cultural associates are not critical, agnotobiotic cultures are perfectly acceptable since they most often require less fastidious treatment and care. Arnold (1954b, 1966, 1967), Lee and Muller (in press), Lee *et al.* (1970) and Myers (1937) have all given detailed directions for the establishment of agnotobiotic cultures of foraminifera. For the sake of completeness, some important hints and suggestions for successful agnotobiotic cultures will be briefly repeated here. Only a few precautions need be observed for success.

Most easily harvested are the species living in the epiphytic communities. Many small samples have to be taken since foraminiferal distribution is so patchy. A hand lens ($\sim$10$\times$) is helpful in locating them in the field. Algae are carefully harvested into plastic buckets, small plastic bottles, or plastic bags. Collections are placed in the shade or in some protective container such as a picnic ice chest. As soon as possible the algae are shaken gently in the collection water and removed. If a large number of foraminifera are being harvested it is sometimes useful to pour the samples through brass sieves (we routinely use #s 4, 18, 35; 4.56, 1.0 and 0.35 mm respectively). Since the metal is toxic to the foraminifera, exposure of the organisms to the brass should be as brief as practical. The pails, bottles, or bags are allowed to stand without agitation for $\sim$15 min in order to allow the foraminifera and diatoms to settle to the bottom. The sea water above the sediment is then decanted and replaced by fresh sea water. Samples are returned to the laboratory in wide mouth plastic screw-capped bottled or plastic bags. It is extremely important to *avoid placing too much sample in any container*. As a general rule we fill our containers with a thin layer of sediment ($\lesssim$20% total volume) overlain with 3$\times$ as much sea water ($\sim$80% of the container is filled). Samples in the field are immediately brought into the laboratory, submerged under the water at the collection site, or placed in an ice chest for transportation to the laboratory. Such procedures avoid the combined detrimental effects on the foraminifera of elevated temperature, rapid bacterial growth, and anoxia.

As soon as possible in the laboratory the samples are placed in dishes which have a large surface area to volume ratio. We use stacking aquaria (Wheaton 41684) but many types of bottles, aquaria, and culture dishes have been successfully used by various workers. Some type of lid should be provided to retard evaporation. The dishes should be incubated in moderate light (250 μW/cm^2) and in a room or chamber which is $\approx$ to the water temperature in which the foraminifera were collected. The spectral quality of the illumination does not seem to be a critical factor for the success of foraminiferal cultures; natural, incandescent, or fluorescent lighting are all useful. As a general rule, temperature of the incubator or room should seldom, if ever, exceed 10° C over the field temperature. Growth and reproduction is usually slower at lower temperatures. As soon as practical, large competitive or predative metazoa should be removed from the initial cultures by pipette or with small forceps.

For the most part, freshly collected epiphytic and psammolittoral foraminifera will crawl through the sediment and collect at the surfaces or on the wall of the container overnight where they can be easily removed by means of Pasteur pipettes and transferred to fresh dishes. Millipore filtered sea water from the collection site is the medium we prefer to use for our crude (=agnotobiotic) cultures but many prefer erdschreiber. Erdschreiber is an enriched sea water medium containing soil extract 5% (V/V), $Na_2HPO_4 \cdot 7H_2O$ (0.005% (W/V) and $NaNO_3$ 0.01% (W/V). The pH of the sea water or erdschreiber should be ~pH 8.0. The medium is changed daily for the first week of culture, changed weekly for the next 3 weeks, and then can be changed less frequently (~monthly) thereafter. Transfers are made whenever the cultures become denser than 10–50 organisms/cm² of bottom surface. In agnotobiotic culture inocula of 50 *R. leei, S. hyalina, Q. lata* yield ~800–1000, 1000–2000 and 3000 animals per month respectively (Muller and Lee, 1969).

2. GNOTOBIOTIC CULTURES

Most of the experiments in our laboratory have required gnotobiotic cultures (all of the organisms present are known). Depending upon the facilities available and the experience of the investigator, there are at least 3 approaches to gnotobiotic culture of foraminifera: (1) fortuitous methods; (2) tracer feeding and related inductive techniques; (3) reductive techniques.

a. Fortuitous Methods

Fortuitous methods are always worth a trial since they require the least amount of effort. As soon as possible after collection, the desired species of foraminifera are picked and separated from the other organisms with the aid of glass needles, fine camel's hair brushes, forceps and Pasteur pipettes. When working with delicate species we have often delayed separation of the foraminifera from the debris and other organisms by 8–12 h. A thin layer of the raw collection is spread out on the bottom of a Pyrex baking dish (18 × 12 × 2½ in; Corning 3170) or in Wheaton (155 × 55 mm; Wheaton 41648) stacking culture dishes. The sediment is overlain with sea water from the collection site and then the dishes are covered with an identical dish. Most epiphytic and epibenthic foraminifera in natural collection spread out this way climb to the surface of the sediment overnight and collect in small balls of algae which they draw together with their pseudopods. The foraminifera are then transferred to the wells of a sterile 9-hole spot plate (Pyrex 7220) kept within a microscope glove box. Sterile sea water with or without mixtures of antibiotics (Lee and Muller, in press, Lee *et al.*, 1970) is aseptically added to the other 8 wells. The foraminifera are transferred by means of Pasteur pipettes into the next well in the smallest practical volume of washing medium. Approximately 20 serial transfers in the course of 4 h will free most foraminifera of externally associated micro-organisms.

After washing, the foraminifera are aseptically inoculated into either liquid or solidified erdschreiber or sea water along with a group of diatoms and chlorophytes, i.e., *Dunaliella parva*, *Dunaliella salina*, *Chlamydomonas* spp, *Nannochloris* sp, *Chlorococcum* sp, *Nitzschia acicularis*, *N. brevirostris*, *Cylindrotheca closterium*, *Phaeodactylum tricornutum*, *Amphora* spp, *Fragilaria construens*, and *Navicula diversistriata*, all found to be good food sources for a variety of foraminiferal species (Lee *et al.*, 1966, Muller and Lee, 1969, Muller, 1972). Gnotobiotic cultures of *Allogromia laticollaris*, *Rosalina leei*, *Spiroloculina hyalina*, *Ammonia beccarii*, *Quinqueloculina lata*, *Bolivina* sp and *Rotaliella* spp have been established this way. Both *Allogromia laticollaris* and *Rosalina leei* grow best with their food in liquid erdschreiber with a salinity of ~26‰ (Muller and Lee, 1969). *Spiroloculina hyalina* grows better on erdschreiber agar slopes with an overlay of sterile filtered (Millipore) sea water. The other species grow better in filtered, sterilized sea water.

b. Inductive Methods

Inductive methods involve elaborate techniques and can take longer, but give greater promise of favorable results, particularly when working with organisms which have not been cultured previously. One of the most powerful approaches is the tracer feeding technique (Lee *et al.*, 1966) which can be used as an effective tool to select the food organism which will most probably support the growth of foraminifera in limited gnotobiotic (synxenic, with few[est] associated organisms) culture. Potential food organisms are grown in ^{32}P labeled media and fed to freshly collected or agnotobiotically cultured organisms. After incubation for a day with radionuclide labeled food the foraminifera are harvested into 9-hole spot plates containing sea water. They are then serially washed to remove adhering food organisms and debris. With hardier species it is possible to harvest the foraminifera into screw-cap test tubes containing sea water. The tubes are then agitated with a vortex mixer. The animals are washed quite well by this method. After washing the foraminifera are counted and placed into vials. Radioactivity is measured by β liquid scintillation spectrometry. Experience with various animals and algae and with various radionuclides over the years has led us to believe that there is less experimental error with ^{32}P than when ^3H, ^{35}S, ^{55}Fe, or ^{14}C are used as labels. Error is greatest with ^{14}C because of label loss during respiration. The level of radionuclide in the potential food organism cultures is adjusted proportionately to their generation times. Our goal is to feed organisms which have a ^{32}P label $\sim$0.5–10 dpm/organism. If the radionuclide label is too low, counting error and experimental errors may become significant as one approaches background levels. If the level of label is much above 10 dpm/food organism and if the food is eaten in relatively large amounts then radiation effects on the foraminifera affect the results. Most of the salt marsh algae with which we have worked have logarithmic growth phases lasting 6–10 days in test tube (10 ml) culture. To obtain the desired level of label we use $\sim$100 μCi ^{32}P/tube. At harvest the radionuclide labeled food organisms are spun down and the medium is decanted. (From this point on aseptic technique is not necessary.) The cells are resuspended in unlabeled media and then spun down again. After 3 serial washes the cells are enumerated, resuspended at a desired concentration, usually 10^5–10^7 cells/ml, and fed to the foraminifera.

We have used another useful inductive method for the isolation of ecologically significant species of nematodes and there is no reason why it should not also be useful in the isolation of limited agnotobiotic cultures of many species of foraminifera since preliminary experiments with *Elphidium incertum* were successful. This auxotrophic method takes advantage of the relative differences in the nutrient requirements of the large variety of bacteria and diatoms growing in the natural epiphytic community (mentioned earlier in the discussion of field studies). Knowledge of the population structure of the foraminifera, algae, and bacteria, and some insight into the nutritional and physiological requirements of the algae and bacteria which peak when the desired species of nematode (foraminifera) blooms in the summer, led to the choice of some selective media (out of the 24 we have developed) which have a high probability of growing the food most likely to support the growth of the particular nematode or foraminifera (Lee *et al.* ms *op. cit.*). It is appropriate to point out at this time that reliance on erdschreiber as a "universal" culture medium for foraminifera will severely restrict the potential types of foraminifera which can be grown in the laboratory. In our recent study, erdschreiber was a poorer medium for the isolation of algae and bacteria than sea water from the local environment alone; $<\frac{1}{2}$ of the total numbers of colonies and colony types grew on erdschreiber than on sea water. Many media had higher or lower medium spectrum indicies (a measure of the ability of the medium to unselectively [or conversely, selectively] support the growth of large numbers of species of algae). Only 20–25% of the total numbers of algal types isolated grew on erdschreiber.

Foraminifera are harvested from natural collections and transferred to 9-hole spot plates or petri plates. Under aseptic conditions they are separated from other microherbivores and inoculated into solidified and liquid varieties of the predetermined kinds of differential media. Incubation is at collection temperature. Depending upon the medium a mixture of antibiotics (dihydrostreptomycin, 2000 μg/100 ml; penicillin 1000 units/100 ml; and Mycostatin 50 μg/ml) may be added to depress bacterial growth. Media are changed 2 or 3 times a week initially but may be changed less frequently if observation of the cultures suggests that this is practical. The algae which grow are usually the species which the foraminifera have gathered around them in the raw collections. Alternatively one can aseptically streak out the raw collection on a series of differential media in petri plates, isolate clones of the most numerous species of algae and bacteria, grow them separately, and feed them to the foraminifera.

c. Reductive Methods

In the end reductive approaches are usually necessary to obtain axenic, monoxenic (with one food species) or dixenic (with 2 food species) cultures. One usually starts with successful laboratory cultures and gradually attempts to eliminate organisms which do not stimulate growth from the culture. Antibiotic mixtures and the aseptic washing techniques mentioned previously (Lee *et al.*, 1970, Lee and Pierce, 1963, Muller and Lee, 1969, Tietjen *et al.*, 1970) are usually successful routes toward this goal. Monoxenic culture of *Allogromia* sp (strain NF) and synxenic cultures of *Spiroloculina hyalina, Allogromia laticollaris,* and *Rosalina leei* were isolated in this manner (Lee and Pierce, 1963, Muller and Lee, 1969, and Muller, 1972). The results of a recent screen of the antibiotic sensitivity of 20 bacterial species isolated from salt marshes are in press (Lee and Muller, in press). Antibiotics can be used either in washes, in the incubation media or both.

Though foraminifera would seem to be able to adapt to wide ranges of physical variables in the salt marsh, there are clear boundaries to be observed in the laboratory if one hopes to obtain high productivity. As a rule of thumb, we usually start our cultures at 25° C, pH 8.1, and 25‰ salinity, which approximate average summer conditions at Towd Point Marsh (Lee *et al.*, 1969a, Matera and Lee, 1972).

B. Niche Related Studies

1. ABIOTIC STUDIES

a. Temperature, Salinity, pH

Quite a number of workers have concerned themselves with the effects of temperature, salinity and pH on the growth of various species of foraminifera in the laboratory. These studies are very important to the interpretation of the potential and realized niches of those species. With respect to each of the variables there are 5 critical ranges of importance: (1) lowest part of the range for survival (incipient); (2) lowest part of the range which permits reproduction; (3) optimum (shortest generation time); (4) highest part of the range which permit reproduction; (5) maximum part of the range for survival (incipient). Among the organisms which have been studied with respect to some or all of these criteria are: *Discorbinopsis aguayoi* (Arnold, 1954); *Spirillina vivipara* (Myers, 1936); *Rotaliella heterocaryotica, R. roscoffensis, Ammonia beccarii, Bolivina compacta, B. vaughani* and *Rosalina columbiensis* (Bradshaw, 1962); *Allogromia* sp (strain NF) (Lee and Pierce, 1963); *Allogromia laticollaris, Rosalina leei* and *Spiroloculina hyalina* (Muller, 1972); and *Quinqueloculina lata* (Muller and Lee, 1969).

b. Multivariate

Valuable as these studies have been as aids in clarifying niche interpretations (summarized as Tables 2–4), there is some hesitation in accepting

TABLE 2

The effect of temperature (°C) on the reproduction of various species of foraminifera cultured in the laboratory

Species	Survival range	Reproductive range			Author
		Lower	Optimum	Upper	
Allogromia					
laticollaris		14	18–29	32	Arnold, 1955
		10		33	Muller, 1972
Spiroloculina					
hyalina		14		24	Arnold, 1964
		10	29–30	33	Muller, 1972
Ammonia beccarii	$\lesssim 0 \sim 45$	20(<15)	30	32	Bradshaw, 1957, 1961
Patellina corrugata		18	21	25	Myers, 1935
Spirillina vivipara	~ 39	18		26	Myers, 1936 Bradshaw, 1961
Tretomphalus sp		18	20		Myers, 1943
Rosalina globularis		16		25	Slitter, 1965
Rosalina leei		10	25	33	Muller, 1972
Bolivina doniezi		18		22	Slitter, 1970
Rotaliella					
heterocaryotica	$\leqq 14.5 \geqq 30$		~ 23		Bradshaw, 1955
Rotaliella					
roscoffensis	$\leqq 14.5 \geqq 30$		~ 23		Bradshaw, 1955
Bolivina compacta	~ 41				Bradshaw, 1961
Massilina sp	~ 40				Bradshaw, 1961
Rosalina					
columbiensis	~ 39				Bradshaw, 1961
Bolivina vaughani	~ 38				Bradshaw, 1961

TABLE 3

The effect of salinity (‰) on the reproduction of
foraminifera cultured in the laboratory

Species	Survival range	Reproductive range			Author
		Lower	Optimum	Upper	
Allogromia laticollaris	>0–10<~60	20	30–40	45	Muller, 1972
Spiroloculina hyalina	~5–≳55	17	20–40	44	Muller, 1972
Rosalina leei	~5–≳60	18	30	47	Muller, 1972
Elphidium crispum	19.5–48 ~25–~35		30–35		Myers, 1943 Murray, 1963
Rotaliella heterocaryotica	16.8–~40	23.5	26–30	37	Bradshaw, 1955
Ammonia beccarii var. tepida	7–>67	13	34	56 40 (Texas)	Bradshaw, 1957 Bradshaw, 1961
Discorbinopsis aguayoi	20–57				Arnold, 1954

TABLE 4

The effect of pH on the reproduction of foraminifera
cultured in the laboratory

Species	Survival range	Reproductive range			Author
		Lower	Optimum	Upper	
Allogromia laticollaris	~3–9.5	5.0	5.5–8.5	9.5	Muller, 1972
Spiroloculina hyalina	~3–8.5	6.0	6.2–8.0	8.0	Muller, 1972, McEnery and Lee, 1970
Rosalina leei	~3–10	6.9	7–9	10.0	Muller, 1972, McEnery and Lee, 1970
Elphidium crispum	<7.3–8.1				Myers, 1943
Ammonia beccarii	3.0–<9.3				Bradshaw, 1957

the data as reflecting the total picture of the restraints placed on the
species throughout its entire range of distribution. Field and laboratory
studies of *Ammonia beccarii* (Boltovskoy, 1964, Boltovskoy and Lena, 1969,
Bradshaw, 1957, Brooks, 1967, Buzas, 1965, 1969), a very widely distributed and often observed species, suggest that there may be physiological

races in various parts of the world. In addition, Bradshaw's (1968) field study and Hutchinsonian approaches to niche theory strongly indicate that multidimensional studies might be more appropriate for interpretations of foraminiferal niches. Muller (1972) recently has made strides in the right direction in his multidimensional study of *Spiroloculina hyalina*, *Allogromia laticollaris* and *Rosalina leei*. The generation times of the 3 species tested were strongly affected by varying several abiotic factors simultaneously (Figs. 8–14).

Illustrative of these kinds of relationships are the synergistic effects on the generation time of *Rosalina leei* by varying both temperature and pH. At 10° C its biotic potential was only 1% of its potential at 30° C

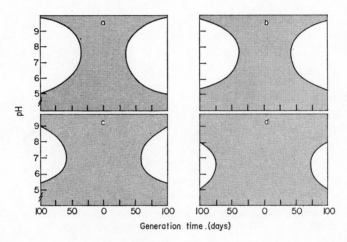

Fig. 8. The effect of pH and temperature on the generation time of *Rosalina leei*; a, upper left, 25° C; b, upper right, 30° C; c, lower left, 15–20° C; d, lower right, 10° C. (Replotted after Muller, 1972).

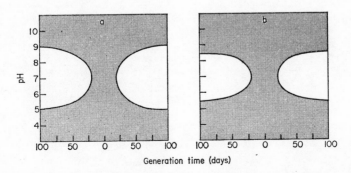

Fig. 9. The effect of temperature and pH on the generation time of *Spiroloculina hyalina*; a, 25–30° C; b, 10–20° C. (Replotted after Muller, 1972).

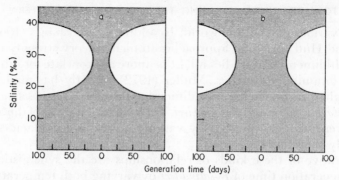

Fig. 10. The effect of salinity and pH on the generation time of *Spiroloculina hyalina*; a, pH 5.5–9.5; b, pH 4.0–5.0 and 10.0. (Replotted after Muller, 1972).

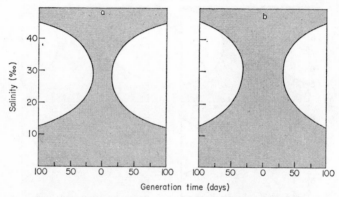

Fig. 11. The effect of temperature and salinity on the generation time of *Spiroloculina hyalina*; a, 20–30° C; b, 10–15° C. (Replotted after Muller, 1972).

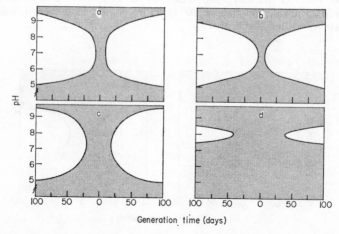

Fig. 12. The effect of temperature and pH on the generation time of *Allogromia laticollaris*; a, 25° C; b, 30° C; c, 15–20° C; d, 10° C. (Replotted after Muller, 1972).

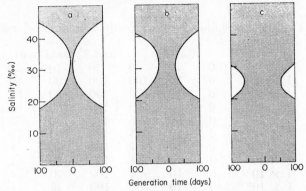

Fig. 13. The effect of temperature and salinity on the generation time of *Allogromia laticollaris*; a, 25° C, b, 20° C, c, 10° C. (Replotted after Muller, 1972).

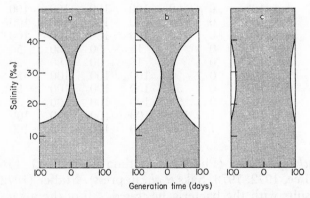

Fig. 14. The effect of pH and salinity on the generation time of *Allogromia laticollaris*; a, pH 7.5–9.5; b, 6.0–7.0; c, 5–5. (Replotted after Muller, 1972).

and reproduction was restricted to a very narrow range around pH 7.0 (Fig. 8). At intermediate temperatures, 20 and 25° C, reproductive capacity was 41 and 72% of those grown at 30° C.

c. Effect on Food

It must be remembered that abiotic variables also affect the growth of food organisms and thus simultaneously affect the food supply of foraminifera if this is not controlled in the experiment (e.g., Table 5). The growth of 4 species of algae which are good food organisms for foraminifera, *Phaeodactylum tricornutum* (strain 39), *Amphora* sp (5), *Nannochloris* sp (41), and *Amphora* sp (RF-8), gradually increased to a maximum at 30° C. Five other species, *Chlamydomonas subehrenbergii* (93), *Chlorococcum* sp (38), *Nitzschia hungarica* (Pb-13), *Cylindrotheca fusiformis* (Bl 27)

TABLE 5

The effect of salinity on the growth of axenic food cells

Algal Strain #	Maximum growth (100%)	Salinity (‰) and % maximum growth							
		0	10	20	30	40	50	60	70
41	1×10^7	0	50	100	100	100	60	1	0
8	1×10^7	0	50	100	100	100	60	1	0
38	1×10^7	0	50	100	100	100	60	1	0
93	5×10^5	0	60	75	100	100	75	1	0
5	5×10^5	0	60	75	100	100	75	1	0
RF-8	1×10^6	0	10	<1	100	40	1	0	0
BL-27	1×10^6	0	10	<1	100	40	1	0	0
Pb-6	1×10^6	0	10	<1	100	40	1	0	0
Pb-13	1×10^6	0	10	<1	100	40	1	0	0
39	5×10^4	1	40	75	90	50	30	10	0
Bacteria Strain #									
D1-701	4×10^8	0	60	100	100	100	100	100	10
A 5-6	4×10^8	0	60	100	100	100	100	100	10
A 5-7	4×10^8	0	60	100	100	100	100	100	10
A5-703	3×10^8	0	2	16	90	80	50	10	0
A1-711	3×10^8	0	2	16	90	80	50	10	0
D2-703	9×10^7	0	50	100	100	100	50	10	0
C1-704	9×10^7	0	50	100	100	100	50	10	0
A3-111	9×10^7	0	50	100	100	100	50	10	0
D 2-7	4×10^7	1	100	100	100	100	100	10	0
D5-702	4×10^7	1	100	100	100	100	100	10	0

and *Nitzschia frustulum* (Pb-6) had optima near 20° C (Muller, 1972, Saks and Lee, 1972, 1973, Saks *et al.*, in press). Muller (1972) obtained similar results with the bacteria he tested. All of the algae tested had upper thermal limits near $\lesssim 38.5°$ C. The most thermal tolerant species, *Amphora* spp (B1-17, RF-8), *Dunaliella quartolecta* (50), *Chlorococcum* sp (38), *Cylindrotheca closterium* (9), *Nitzschia* sp (B1-15 A), survived only 135–195 min when subjected to acute thermal stress at 40° C (Saks and Lee, 1972, 1973).

d. Dynamic Aspects

Very few studies have attempted to interpret the dynamic aspects of abiotic stresses. Phleger and Bradshaw (1966) for instance drew attention to the daily pH shifts in the Mission Bay, Calif. marsh they studied. During the night pH was low ($\sim$6.8) and during the day it rose ($\sim$8.5). They found that the pH of the marsh was below 7.6 $\sim$ half of the time, during which the foraminiferal shells would tend to dissolve unless the organisms expended energy to maintain them. Angell (1967) decalcified the tests of both *R. floridana* and *S. hyalina* by immersing them in acidified sea water (pH 5.0; adjusted by adding HCl) for less than 3 h. Although some damage was done to the organisms by the treatment, i.e., collapse of the terminal chamber, detachment of the terminal chambers, blebbing of the cytoplasm, most of the organisms resumed feeding after 3 h and *R. floridana* produced viable offspring. *S. hyalina* did not recalcify and eventually degenerated but *R. floridana* recalcified, somewhat abnormally, after 3–5 days following rapid decalcification. McEnery and Lee (1970) were more successful with gentler methods of decalcification. They used autoclaved sea water adjusted to pH 8.0 with 0.1 M citric acid or 0.1% (W/V) Na_2EDTA in sea water. Removal was most rapid when the decalcification medium was changed every h for a period of 24–48 h.

Although both species of foraminifera completely recovered and there was no evidence of permanent damage, the process of schizogony in treated *Rosalina* was delayed $\sim$2 weeks past that of the controls. Many specimens fed actively during the decalcification process.

2. BIOTIC STUDIES

a. Food

Many earlier workers made qualitative observations on feeding and the food of foraminifera (Føyn, 1936, Jepps, 1926, LeCalvez, 1938, Lister, 1895, Myers, 1935, 1936, 1942a, b, 1943 and Sandon, 1932, etc.). Diatoms were the most easily recognized and most frequently reported food organisms although small flagellates, macrophyte gametes, small eggs, other protozoa, unicellular and filamentous algae, bacteria, faecal pellets, and micrometazoa have been reported as food. Myers, in his study of *Elphidium crispum*, was able to make inferences on the general influence of food (phytoplankton abundance) upon the growth and reproduction of this species in Plymouth Sound, Villefranche-sur-mer, La Jolla, Calif., and Poela Kelapa in the Thousand Islands near Sundra Strait in the Java Sea.

i. Growing Foraminifera with Food

Though microscopic observations of freshly collected specimens or examination of successful agnotobiotic cultures led to some understanding of foraminiferal nutrition and were important to the advancement of culture techniques (Myers, 1937, Arnold, 1954) only recently has this aspect of foraminiferal biology been explored systematically. Various investigators have used different methods in the assessment of nutritional value. Each of the methods, however, is not without serious drawbacks. Perhaps the simplest and seemingly most straightforward approach is to inoculate different species of food into cultures of the test animals and to measure their growth and reproduction. One dilemma created by the method is that caused by algal metabolites which may build up in the culture.

Release of soluble organic compounds by algae has been shown in almost every careful study of this problem (Fogg, 1966). Some algal symbionts can release as much as 85% of their photosynthetically fixed carbon (Smith *et al.*, 1969) but most algae release only a fraction (∼3–6%; Antia *et al.*, 1963, Fogg, 1971, Hellebust, 1965, Horne *et al.*, 1969, Samuels *et al.*, 1971). Natural populations tested by Hellebust (1965) released ∼4–16% of their photoassimilated carbon during rapid growth and ∼17–38% at the end of their blooms when some cells may be dead or dying. He also found that some species (i.e., *Dunaliella tertiolecta*) released much more photosynthetate at natural light levels than at lower levels. The released compounds include many types of substances: proteins; peptides; polysaccharides; monosaccharides; polyols; amino acids; glycolic acid; and various vitamins (Carlucci and Bowes, 1970a, b, Fogg, 1971, Hellebust, 1965). Algal and foraminiferal waste products have been shown to inhibit the reproduction of *R. leei* in some bacterized gnotobiotic cultures and to enhance growth in others either directly or by their effect on the other microflora and bacteria present in the cultures (Muller and Lee, 1969). Growth of *R. leei* was 3 times greater in cultures in which a mixture of the diatoms *Nitzschia acicularis*, *Cylindrotheca closterium*, *Phaeodactylum tricornutum*, *Amphora* sp, and *Nitzschia brevirostris*, was washed by centrifugation than when it was not. Growth on an unwashed mixture of chlorophytes, *Chlorococcum* sp, *Chlamydomonas* sp (Ar-s) and an unidentified chlorophyte (SH-1), was ∼ the same as growth on the unwashed diatoms. Growth was slightly better with washed chlorophycean cells. When media were changed weekly some of the relationships were changed. *Rosalina* grew better (2×) in the presence of washed chlorophytes than in washed diatoms. Eventually the unwashed diatom cells caught up with the chlorophyceans.

ii. Heat Killed Food

One plan to avoid the effects of algal metabolites is to kill the food organisms. Bradshaw (1961) found that *Rotaliella heterocaryotica* ate both living and heat killed *Chlamydomonas* sp. Final foraminifera populations were higher when the organisms were fed twice as many *Chlamydomonas*. When *Nitzschia* sp was the food organism populations reached only 35% of the levels of *Chlamydomonas*-fed organisms. *Rotaliella roscoffensis* on the other hand, did not grow on dead *Chlamydomonas* but grew well on the same species of *Nitzschia* which supported poor growth of *R. heterocaryotica*. The growth responses of *Ammonia beccarii* var. *tepida* to *Dunaliella* sp were similar to those of *Rotaliella heterocaryotica*; growth was proportional to the concentration of the food organisms fed (Bradshaw, 1961). Food concentrations below 220000 cells/culture did not permit any growth or reproduction. Bradshaw used antibiotics (penicillin 72 U/ml and streptomycin 83 μg/ml) to suppress bacterial growth which might otherwise have occurred when the food decomposed. Murray (1963) used an entirely different approach in his study of *Elphidium crispum*. His experiments were shorter in duration and he used the daily rate of feeding cyst formation to measure feeding rates. He killed *Phaeodactylum tricornutum* by irradiating cultures under an ultraviolet lamp for $\frac{3}{4}$ hr or by heating them at 80° C for 20 min. *Elphidium crispum* did not eat food killed by either method. Murray designed an appropriately controlled experiment in order to test whether the killing processes made the food unacceptable to the foraminifera or whether living *Phaeodactylum tricornutum* produced a substance which encouraged *Elphidium* to feed. His protocol included a control with living *Phaeodactylum* in their culture fluid and another one with dead *Phaeodactylum* in their culture medium. Experimental cultures were given either an equal mixture of living or dead cells, a mixture of dead cells plus fluid from living cells, a mixture of living *Phaeodactylum* plus fluid from irradiated cultures, or dead cells alone. Most feeding cysts were formed by foraminifera in the living controls. The presence of dead cells or culture fluid from irradiated cells lowered feeding rates but some dead cells were eaten by the animals. No evidence was obtained which could be used to suggest that an appetite teasing substance was released by the algae.

iii. Tracer Feeding Studies

A different approach was taken by Lee and his co-workers (Lee *et al.*, 1966, Muller, 1972). They used tracer feeding techniques to obtain answers to nutritional questions. The method has the advantage of being rapid and simple to use. It proved to be a real bottle-neck breaker since they were able to test in their initial study more than 50 species of microorganisms including bacteria, yeasts, dinoflagellates, chlorophytes, chrysophytes, diatoms and cyanophytes as potential food for *Allogromia* sp (NF), *A. laticollaris*, *Ammonia beccarii*, *Quinqueloculina* spp, *Rosalina leei*, *Anomalina* sp, *Elphidium incertum*, *Spiroloculina hyalina*, *Globigerina bulloides* and *Globorotalia truncatulinoides*. Contrary to the well accepted views of Sandon (1932) the foraminifera they studied were quite selective in their feeding habit. In general, selected species of diatoms, chlorophytes and bacteria were eaten in large quantity whereas most species of bacteria, cyanophytes, dinoflagellates, and chrysophytes which were tested were not eaten. Of the 28 algal and bacterial species fed to *Rosalina leei*, *Allogromia laticollaris*, *Sprioloculina hyalina*, and *Ammonia beccarii* by Muller (1972) only 4–5 were consumed in significant quantities (0.4–1.5 × 10^{-3} mg/foram/day) (Figs. 15–18). Two species of algae, *Phaeodactylum tricornutum* and *Amphora* sp (5) were eaten in large amounts by all the species tested. Although large numbers of bacteria were eaten, their biomass (0.1–8.3 × 10^{-6} mg/animal/day) was negligible when compared to the algae. They (Lee *et al.*, 1966) also confirmed and expanded Bradshaw's (1961) observations on the effect of food concentration on

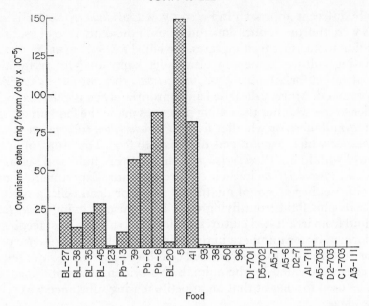

Fig. 15. Feeding rate of *Allogromia laticollaris* on selected species of algae and bacteria. Most organisms were from the Towd Point Marsh; three species of algae, Pb-13, Pb-6 and Pb-8 were from Plum Beach, Jamaica Bay, N. Y. Those organisms on the left of the graph, B1 27-96, are algae; those on the right, D1-701-A 3-11, are bacteria. Names of the organisms given in Muller and Lee (1969) and Muller (1972). (Figure after Muller, 1972).

growth. Feeding behavior by foraminifera on most food was erratic below a threshold concentration ($\sim 10^3$ cells/10 ml culture) and was $\sim$ directly proportional to concentration within a range of 10^4–10^6 organisms/10 ml experimental tube (Fig. 19). Tracer experiments also gave some insight into other aspects of feeding behavior. Small *Allogromia laticollaris* (~ 150–200 μm in diameter) ate many more food organisms than did larger (~ 350–400 μm) ones. They also grow much faster (Lee *et al.*, 1969b). Sometimes the difference was as great as 200%. Similar results were also reported for small and large (young and old) *Allogromia* sp (NF). Specimens of *A. laticollaris*, *Allogromia* sp (NF), *R. leei* and *A. beccarii*, the most extensively studied species, also varied in their feeding behavioral responses to the growth phase characteristics of some species of algae. At one extreme *A. beccarii* ate almost 5× more log phase *Chlorococcum* sp than cells from stationary phase cultures. One of the limitations of the tracer feeding method as it applies to foraminifera (Lee *et al.*, 1966) is that it does not distinguish between ingestion or food gathering, and assimilation. Foraminifera often gather huge balls of food which are digested sometime later.

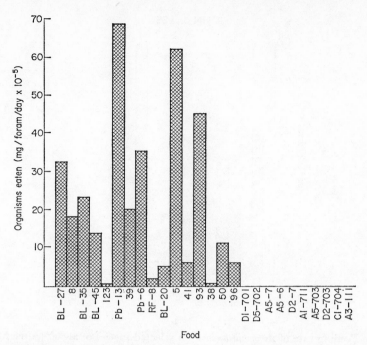

Fig. 16. Feeding rate of *Rosalina leei* on selected species of algae and bacteria. (After Muller, 1972).

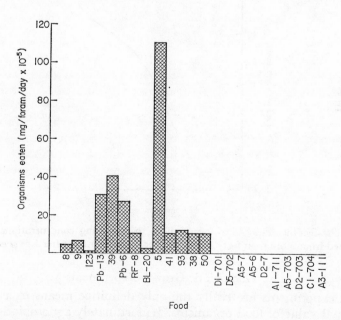

Fig. 17. Feeding rate of *Spiroloculina hyalina* on selected species of algae and bacteria. (After Muller, 1972).

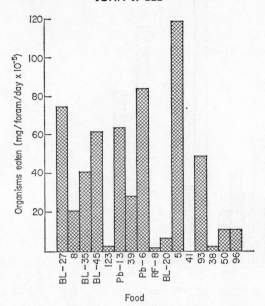

Fig. 18. Feeding rate of *Ammonia beccarii* on selected species of algae. (After Muller, 1972).

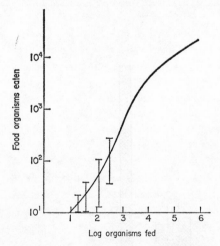

Fig. 19. The feeding response of foraminifera to increasing concentrations of food. Generalized from more than forty tracer feeding experiments. (After Lee *et al.*, 1966).

iv. Gnotobiotic Growth Experiments

Growth experiments are really the only definitive means to assay the nutritional value of food organisms. Unfortunately agnotobiotic nutritional experiments yield only circumstantial evidence. Only one foraminifer, *Allogromia* sp (NF), has yet been grown in continuous monoxenic culture (Lee and Pierce, 1963). None have been isolated axenically, and only a few, *Allogromia laticollaris, Spiroloculina hyalina* and *Rosalina leei*, have been raised in synxenic culture (Lee *et al.*, 1963, Muller and Lee, 1969). Some types of nutritional questions can be answered with synxenic cultures; others require axenic cultures to obtain definitive answers uncomplicated by the relationship between foraminifera, their food, and the substrate. Foraminifera in synxenic culture require a metabolite or metabolites, essential for their rapid growth and reproduction which are found only in selected species of bacteria or in rate limiting quantities in the algae which have thus far been examined (Muller and Lee, 1969, Lee and Pierce, 1963). Suppression of bacterial growth by the use of antibiotics or reduction or elimination of them by aseptic washing techniques results in dramatic lowering of fecundity. Higher levels of bacteria ($\geqq 1 \times 10^6$/ml) are also inhibitory to laboratory cultured foraminifera. Carbon budgets have yet to be drawn up for the relative contributions of bacterial or algal biomass to the growth of any species of foraminifera but there are hints in the literature (Lee *et al.*, 1963, Muller and Lee, 1969, Muller, 1972) based on gnotobiotic studies, that algae, with one possible exception (*Allogromia* sp [NF]) contribute more than do bacteria.

With due respect to all the qualifications just discussed, there is a body of evidence based on gnotobiotic studies of littoral benthic foraminifera which suggests that these animals are highly selective feeders and that they gain different nutritional values from the food that they do assimilate. Mixtures of food organisms sometimes are necessary for growth and reproduction. The nutritional requirements of gnotobiotic clones (synxenic with 2 species of bacteria) of *Allogromia laticollaris* have been studied in detail (Lee *et al.*, 1969b, Lee and McEnery, 1970, Muller and Lee, 1969, Muller, 1972). Animals from the Cold Spring Harbor strain were serially transferred on the same diet in order to deplete nutritional reserves and to intensify the differences caused by the diet. Control cultures grown on agar substrates with only 2 species of bacteria reproduced very slowly by binary fission or cytotomy during 70 days of observation. A large organism produced 5 daughter animals by cytotomy after 21 days; a medium sized organism produced only 2. During the same approximate time medium sized animals grown on *Nitzschia acicularis* produced 53, on *Cylindrotheca closterium* 40, on *Chlorococcum* sp 51, on *Nannochloris* sp 75, on an unidentified chlorophyte (B1-1) 89, on a mixture of *N. acicularis, C. closterium* and *Phaeodactylum* 119, on a mixture of *N. acicularis, C. closterium* and *Chlorococcum* sp 150, and on a mixture of *Chlorococcum* sp, *N. acicularis* and *Nannochloris* 160, all by schizogony. Clone cultures of *Allogromia laticollaris* (CSH strain) did not reproduce continuously on monalgal diets of *C. closterium, Chlorococcum* sp, *Nannochloris* sp, *P. tricornutum*, or on the first 2 mixtures listed above but were fecund on diets of *N. acicularis* alone or with *Chlorococcum* sp and *Nanno-chloris* (Lee and McEnery, 1970). In contrast monalgal cultures of either *Chlorococcum* sp or *Nannochloris* supported the continuous reproduction of the Towd Point strain of this species. Mixtures of these algae plus *N. acicularis* or *C. closterium* also supported continuous reproduction. The nutritional requirements of both strains were so dissimilar that the only experimental diet which satisfied the growth of both strains was a mixture of *N. acicularis*, *Chlorococcum* sp and *Nannochloris* sp (Lee and McEnery, 1970).

Similar results were obtained with *Spiroloculina hyalina*, *Quinqueloculina lata* and *R. leei* (Muller and Lee, 1969). Gnotobiotic bacterized (2 species) *R. leei* reproduced quite rapidly on a diet of *Monochrysis lutheri*, *Chlamydomonas hedleyi*, *Isochrysis galbana* and *Nitzschia brevirostris* fed singly or in combination. Gnotobiotic bacterized cultures of *Q. lata* reproduced every 3 weeks when grown gnotobiotically with either *Chlorococcum* sp or a mixture of *N. acicularis* and *C. closterium*. *S. hyalina* had very similar requirements. Gnotobiotic cultures grew well when *C. closterium*, *Nitzschia* sp (B1-15A and B1-17), *D. parva* or *Chlorococcum* were added singly or in mixtures. Growth of *S. hyalina* was generally $4 \times$ greater on the mixtures than in the monalgal cultures. They found, however, that gnotobiotic cultures of *Quinqueloculina lata* and *Q. seminulum* grew and reproduced $4 \times$ faster in a mixture of *Chlorococcum* sp, *C. closterium* and *Phaeodactylum tricornutum* than they did in agnotobiotic stock cultures. This suggests that there are upper bounds to the enhancement of growth by increasing the numbers of algal species.

Though the dietary habits of foraminifera seem quite varied, the behavior of one motile algal species, *Dunaliella* spp, to many species of foraminifera is interesting. *Dunaliella* cultures were observed by Lee and his coworkers (1961) to stimulate rapid pseudopodial formation in starved cultures of *Globorotalia*, *Orbulina* and *Quinqueloculina lata*. When specimens of *Q. lata*, *Ammonia beccarii*, *Bolivina* sp and *Trochammina inflata* were placed in dishes containing *Dunaliella parva*, a large proportion of the flagellates changed from random locomotion and swam toward the test or the pseudopodial net. When they reached the net or the test they stopped swimming and eventually lysed. This Circean effect may be a general one. Belar (1923, 1924) observed a similar phenomenon in cultures of *Actinophrys sol*.

b. Density Dependent Interactions

i. Intraspecific

Very little is known about the other biotic aspects of foraminiferal niches. Muller (1972) measured the carrying capacity and the generation time of the 3 species of foraminifera he studied. *Allogromia laticollaris* had the highest intrinsic rate of increase ($r = 2.533$ org/day) of those tested.

Under the experimental conditions he used, the carrying capacity was ~2600/10 ml of media which was reached in ~40 days. Under the same experimental conditions *R. leei* had a low intrinsic rate of increase ($r = 0.272$ org/day) and it took almost 2 years (540 days) to reach carrying capacity (1350 org). The intrinsic rate of increase of *S. hyalina* was intermediate ($r = 1.472$) and it reached carrying capacity (3600/10 ml of media) in ~50 days.

Crowding (intraspecific competition) more strongly affected the feeding and rate of reproduction of *A. laticollaris* than it did *R. leei* or *S. hyalina*. A clone culture produced 400% more animals than an inoculum of 10 (Fig. 20). An inoculum size of 33 cut reproduction to ~7% of a

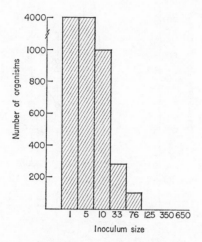

Fig. 20. The effect of crowding on the reproduction of *Allogromia laticollaris*. Incubation time = 100 days. Net number of organisms above inoculum at harvest is plotted. (After Muller, 1972).

clone. Tracer feeding experiments showed great competition for food or interference with feeding in moderately crowded cultures (100 animals). The feeding rate of *A. laticollaris* on *Nitzschia acicularis* and *Phaeodactylum tricornutum* was <30% of the rate of single animals (Fig. 21). In contrast individual *R. leei* in moderately crowded inocula (20–50 animals) ate twice as many algae as did inocula of 5 animals (Muller, 1972). The feeding of *S. hyalina* was not affected by crowding even though the reproductive rate was.

ii. Interspecific

The effects of interspecific competition among different foraminiferal species and among foraminifera and other protozoa and micrometazoa are largely unknown. Initial studies in our laboratory (Lee *et al.*, 1966, and in press, Muller and Lee, 1969, Muller, 1972, Saks *et al.*, in press) indicate that this is a very important factor to consider when interpreting the realized niches of foraminifera. Because so many variables are involved, there appears to be no way in which one can predict the outcome of competitive experiments. Concentration of food, food quality, medium, relative numbers of the test species, and their intrinsic rates of increase all seem to play a role in competitive interactions. *Allogromia laticollaris* and *Allogromia* sp (NF) generally did not interact at low

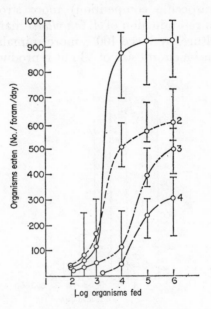

Fig. 21. The effect of intraspecific competition on feeding by *Allogromia laticollaris* on *Nitzschia acicularis* and *Phaeodactylum tricornutum*. Curve (1) inoculum was 1 animal; (2) 10 animals; (3) 50 animals; (4) 100 animals. (After Muller, 1972).

concentrations of food or when present in a ratio of 1:1. At a ratio of 5 *Allogromia* sp (NF): 1 *A. laticollaris* feeding of both species was generally depressed (10–90%, depending on the food; Lee *et al.*, 1966). *R. leei* could not compete with *Allogromia laticollaris* in medium ASP_7 but was able to do so in erdschreiber. *Allogromia* reproduced rapidly to 100–500 fold within 2 weeks. After 4 weeks, *R. leei* began to reproduce and *Allogromia* decreased in number (Muller and Lee, 1969). *R. leei* affects feeding of *A. laticollaris*; in its presence *Allogromia* are only 39% of the *Amphora* sp (5), 23% of the *Amphora* sp (RF-8), 7% of the *Phaeodactylum tricornutum* and 86% of the *Nannochloris* sp (41) it would eat when incubated alone (Muller, 1972). *Spiroloculina hyalina* ate respectively 39, 10, 33 and 23% of the same foods in the presence of *A. laticollaris*. Muller (1972 and ms submitted for publication) was able to develop an equation to reduce this type of information on interspecific competition to a single vector form in order to be able to use it in his graphic analysis of the niche breadth of the 4 species he studied.

$$\text{I.S.C.} = \frac{F_1 + F_2}{2F_a}$$

This equation is an expression of the interspecific competition coefficient.

I.S.C. = the interspecific competition coefficient

F_1 = feeding rate in competition with another species (1)

F_2 = feeding rate in competition with another species, different from (1)

F_a = feeding rate when the species under analysis grows without competition

I.S.C. varies from 0.0 (interspecific competition severe) to 1.0 (no interspecific competition).

c. Niche Representation

His polygonal graphic method is convenient because it gives the viewer a meaningful comparative representation of various niche vectors in an easily interpreted form. He proposed 2 alternative methods of plotting based on the relative weight given to abiotic niche vectors (Figs. 22–26). Using his eccentric model (Fig. 26) *A. beccarii* has a niche breadth of 44.05% meaning that on the basis of laboratory data it is able to occupy (theoretical niche) 44.05% of the total niche area. *Spiroloculina hyalina*

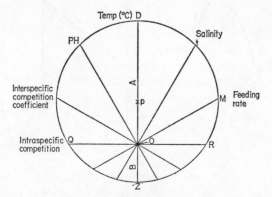

Fig. 22. Graphic representation of 6 factors judged as important determinants of niche breadth in foraminifera. In this representation the center of the plot (P) has been displaced from the center of the circle along the temperature line (to 0) in order to emphasize the traditionally important 3 abiotic factors, pH, salinity and temperature, being plotted. (From Muller, 1972; an alternative method of plot was also developed, Muller ms submitted for publication).

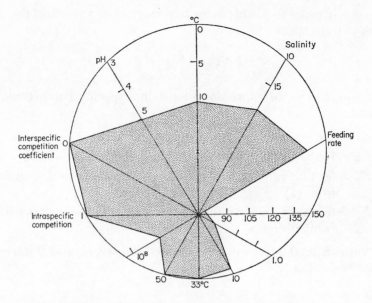

Fig. 23. Graphic representation of the niche of *Allogromia laticollaris*. (From Muller, 1972).

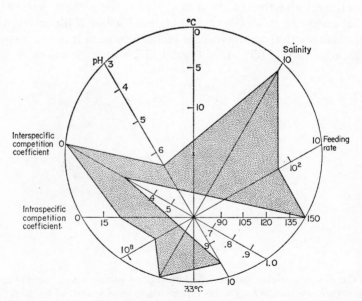

Fig. 24. Graphic representation of the niche of *Spiroloculina hyalina*. (From Muller, 1972).

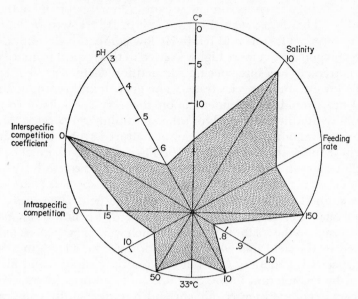

Fig. 25. Graphic representation of the niche of *Rosalina leei*. (From Muller, 1972).

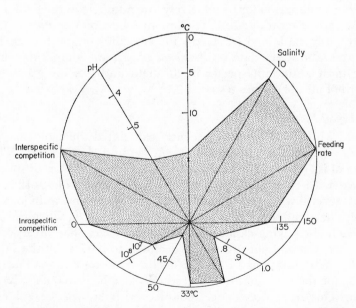

Fig. 26. Graphic representation of the niche of *Ammonia beccarii* (adapted from Bradshaw, 1957, Muller, 1972).

(Fig. 24) had the narrowest niche breadth (26.69%) of any of the forms
he studied. The *A. laticollaris* polygon (Fig. 23) is skewed by high tolerances to physical factors and relatively low competitive abilities. The *S.
hyalina* polygon is even more highly skewed in this respect primarily based
on its extremely low interspecific competition coefficient (0.25), which
was the lowest of the species tested. The growth and reproduction of *S.
hyalina* are probably dependent upon the very specialized conditions
which may inhibit the growth of other foraminiferal species. Although
it eats small amounts of algae it can be cultured synxenically with only
bacteria (Muller and Lee, 1969). *R. leei* has a more regular niche polygon
(Fig. 25), suggesting to the observer that the species is relatively unaffec-
ted by crowding or by competition with other species. None of the 3
species studied by Muller is dominant in the field (Lee *et al.*, 1969a,
Matera and Lee, 1972). Small numbers of *R. leei* and *S. hyalina* were
found in each of the 3 summer seasons which were surveyed. *A. laticollaris*
bloomed in the field following 5 years of drought, suggesting that this
hardy species is generally rare but can multiply rapidly and fill niches
vacated by less tolerant forms. Since *R. leei* has a high interspecific and
intraspecific competition coefficient and has a generally wide range of
tolerance to temperature, salinity and pH, it should be found fairly
abundantly in the field. However, it is only found in small numbers. One
plausible explanation for its relatively low numbers in the field might be
its slow rate of reproduction. The low numbers of *S. hyalina* in the field
are probably explained by its low interspecific competition coefficient.
Field and laboratory data on *Ammonia beccarii* are comparable; it is one
of the most abundant species in the field and has the greatest niche
breadth of all the species tested.

d. Ecological Efficiency

Lee and Muller (1973) have recently calculated the ecological growth
efficiency ($E_e = P/I$) for *S. hyalina*, *R. leei* and *A. laticollaris*. The con-
version of ingested carbon into body tissue was small for all species and
declined with time. *Allogromia laticollaris* had an E_e of 10–12% during
the first several weeks of culture when the population was in logarithmic
growth phase; after 7 weeks the E_e declined to $<1\%$ when the population reached carrying capacity. *Spiroloculina hyalina* had an E_e of 5.19%
in the first week of culture which declined to 0.07% by the 7th week.
The E_e of populations of *R. leei* declined more slowly. The E_e was
21.58% for the first month and gradually declined to 0.83% in the 9th
month. This line of evidence, coupled with knowledge gained from
tracer feeding experiments, standing crops and seasonal succession of
foraminifera, algae, and bacteria suggest that the salt marsh foraminifera
which have been studied in detail are very opportunistic and well
adapted to play brief and specialized roles in the rapidly changing food

webs of the communities of which they are an important part. If the other microorganisms and micrometazoa in the community are found to be similarly adapted then it would seem reasonable to suggest that the high productivity of salt marsh aufwuchs communities is achieved through rapid cycling of very specialized species.

IV. Symbiosis in Foraminifera

The symbiotic association of algae and marine invertebrates is wide spread and has been the subject of much recent research. The question of similar associations occurring in foraminifera, however, has been neglected since its discovery 87 years ago (Butschli, 1886). Indeed Hedley (1964) in the most recent comprehensive review of the biology of foraminifera, questioned the validity of the reports of symbiosis in foraminifera by earlier workers (Butschli, 1886, Doyle and Doyle, 1940, Rhumbler, 1911, Winter, 1907) on the basis that much of their earlier evidence was circumstantial by today's standards. Hedley's doubts stimulated a flurry of research activity on the subject.

A. Symbionts in Planktonic Foraminifera

Lee and coworkers (1965) described the zooxanthellae from *Globigerinoides ruber* and one his students (Zucker, 1973) recently completed a study of their fine structure. The *Globigerinoides* symbionts, possibly a *Merodinium* sp, because of the small size of their motile forms, bear some superficial resemblance to the symbiont *Symbiodinium microadriaticum*, which lives in corals, anemones, and "upside down" jellyfish like *Cassiopea* (Freudenthal, 1962, Kevin *et al.*, 1969, Taylor, 1968). Taylor has suggested transferring this species from *Symbiodinium* to *Gymnodinium*. Arguments could be made on either side of the question, which is really outside the province of this review. They are quite different organisms, however. In *Symbiodinium* the epicone and hypocone are approximately equal in size but in *Merodinium* the epicone is about twice the length of the antapically tapered hypocone. The girdle in *Merodinium* sp has a torsion which leads to a displacement of its ends approximately $\frac{2}{10}$ of its body length. Its sulcus is very shallow and not as well developed as in *Symbiodinium*. The same zooxanthella from *Globigerinoides ruber* was also found in *Globigerinoides conglobatus*, *Globigerinita glutinata* and *Globigerina bulloides* (Zucker, 1973). It is similar but not identical to the symbiont *Merodinium brandti*, found by Hovasse (1923) in the radiolarian *Collozoum inerme*. In contrast to the *Globigerinoides* symbionts the epicone and hypocone of *M. brandti* zoospores are nearly equal in length. The zooxanthellae from *Orbitolites duplex*, a soritid foraminifer studied by Doyle and Doyle (1940) seem, on the other hand, very similar to *Symbiodinium microadriaticum*.

B. Symbionts in Benthic Foraminifera

Because benthic foraminifera are more easily accessible and seem, at present, more amenable to laboratory study, advances in our knowledge of symbiosis in this group of animals has proceded much more rapidly. Recent studies by Rottger (1972a, b) and Rottger and Berger (1972) on *Heterostegina depressa*, a modern nummulitid, have demonstrated enhancement of growth at optimum light conditions (300 lux). Rottger has observed that each of the 100 or more schizozoites receives symbionts upon division of the parent. Their (Rottger and Berger's) other conclusions on the role of the symbionts in the nutrition and growth of the foraminifera are only conditionally valid since their experimental animals were agnotobiotic. Bacteria can and do form an important part of the diet of many species of foraminifera (Muller and Lee, 1970). Critical tracer experiments could clarify the role of the symbionts in the overall carbon budget of *Heterostegina*. The symbionts themselves appear to be diatoms (Dietz-Elbrachter, 1971). Some simple spectrophotometric measurements and examination by standard diatom methods (which would include frustrule cleaning) could easily clarify the taxonomic identity of the *Heterostegina* symbionts. *Marginopora vertebralis*, a large miliolacean, also contains symbiotic algae. Ross (1972) was able to obtain some circumstantial evidence that the symbiotic zooxanthellae might be the same as the *Gymnodinium obesum* or *G. rotundatum* swimming in laboratory cultures near the foraminifera.

Axenic isolation of the zoochlorella from the foraminifer *Archaias angulatus* (Lee and Zucker, 1969) provided the first opportunity to study a foraminiferal symbiont in detail. The organism they studied bore some resemblance to the symbionts from another foraminifer, *Peneroplis*, which Winter (1907) had named *Cryptomonas schaudinni*. The symbiont from *Archaias* was clearly not a *Cryptomonas* as the genus is currently, or commonly in the distant past, characterized. Though both symbionts are green, ovoid, biflagellate, and have a truncated apex, and a prominent anterior vacuole, several important differences were recognized. The nucleus in the *Archaias* symbiont is larger than the one in *C. schaudinni*. The chloroplast is longer and more slender in the *Archaias* zoochlorella. A fairly large stigma was also observed. Because of the small size of the organism, Lee and Zucker (1969) reserved judgement on the taxonomic identity of the *Archaias* zoochlorella until fine structural studies could be done. During the course of nutritional experiments Lee and Zucker (1969) noticed that the puzzling large lipid vacuoles found in the symbionts within their hosts and formed in the culture were no longer formed in certain of their experimental media. The chloroplasts were typical of the volvocales with pyrenoids surrounded by iodine stainable starch. Fine structural studies of the *Archaias* symbiont led to a detailed

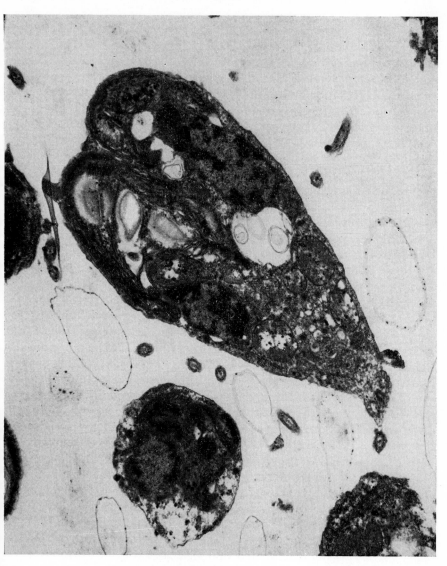

Fig. 27. Dividing motile form of *Chlamydomonas hedleyi*, the zoochlorella isolated from *Archaias*. (From a study by Lee, Crockett, Hagen and Stone, Ms submitted). × 1875.

description and clarified its taxonomic status (Lee *et al.*, in press). It is a typical species of *Chlamydomonas* in most nutritional situations (Figs. 27, 28). However, cells filled with lipid vacuoles are found in the host. Non-motile cells grown in medium "Q_2" resemble those isolated from their host animal. Large quantities of stored lipids characterize such cells. Lipid production involves a projecting type of pyrenoid rather than an embedded type (Lee *et al.* ms *op. cit.*). The pyrenoid seems to be a greatly extended evagination of the outer membrane of the chloroplast (Fig. 29). The cytoplasm in the vicinity of the forming lipid vacuoles is filled with small vacuoles which appear to be formed at the edges of active golgi complexes located in close juxtaposition to the chloroplast.

C. Biology of Chlamydomonas hedleyi

They also found that *Chlamydomonas hedleyi*, the name they gave to the symbiont from *Archaias*, has no vitamin or organic requirements for growth. However, growth is tripled in the presence of thiamine and doubled in the presence of 1 μM glutamic acid, histidine or methionine. Urea was the best nitrogen source tested. Purines and pyrimidines did not serve as nitrogen sources. Optimum phosphate concentration was 0.5 μM. *Chlamydomonas hedleyi* grew well in a salinity range of 6–>52‰ and a pH range of 6–8.5 (Lee *et al.* ms *op. cit.*).

Primary production was estimated both manometrically and with the aid of a radionuclide tracer. The estimates were in close agreement with each other. Approximately 7×10^{-7} M carbon was fixed per gram dry weight of cells per hour. The rate of release of labeled soluble organic material by the axenic symbionts in culture was quite high. Close to the end of logarithmic growth (10 days) more labeled organic carbon (57%) was found in the medium than in the cells. After filtering out of the cells the inorganic carbonate was removed by precipitation after the addition of a saturated solution of $BaCl_2$. The culture fluid was concentrated by flash evaporation, desalted, and chromatographed by thin layer techniques. The organic material concentrated from the medium was chromatographically homogeneous. The R_f values of the compound chromatographed in an ethyl acetate:pyridine:water (8:8:4 by volume) or phenol:water (4:1) or n-butanol:acetic acid:water (4:1:4), propanol:ethyl acetate:water (7:1:2), or n-butanol:ethyl alcohol: water (4:5:1) systems indicated that mannitol was most probably the substance released by the symbiont into the medium (Lee *et al.* ms *op. cit.*).

D. Speculation on the Origin of Symbiosis in Foraminifera

It is easy to imagine how symbiosis can be initiated and established in foraminifera. Many shallow water foraminifera are bloom feeders (Lee *et al.*, 1966, Lee and Muller, 1973). When appropriate species of food organisms are present in sufficient concentration the foraminifera

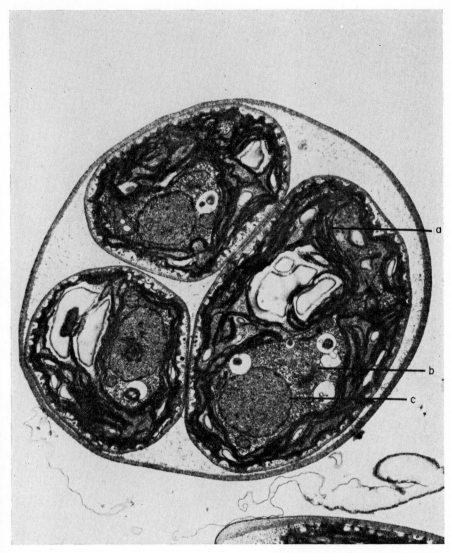

Fig. 28. Typical vegetative cells of *C. hedleyi* in unenriched culture medium. (a) chloro-plast, (b) golgi apparatus, (c) nucleus. (From the above study) ×2880.

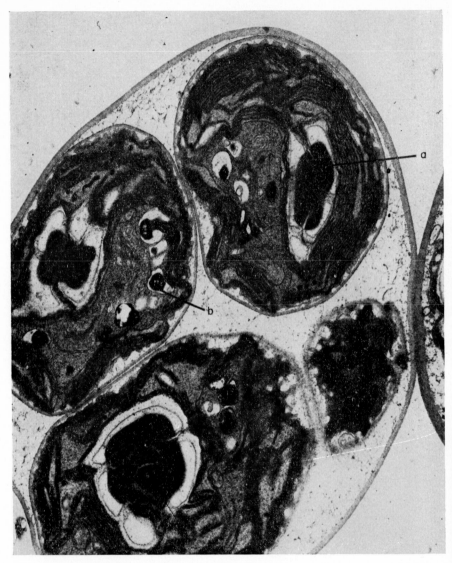

Fig. 29. *C. hedleyi* in enriched culture media. Note 2 types of storage material being formed, starch and lipids. (a) starch, (b) lipid. Lipid storage is characteristic of the symbionts in their host (Lee and Zucker, 1969). × 2880.

capture them and bring them into their tests. Some species pack as many algae as will fit into their newest formed chambers which are otherwise highly vacuolated. Other species gather high masses of food about the external periphery of their shells. At times the foraminifera appear to be embedded in masses of food organisms. Not all of the food is digested at once. In a sense the gathering and hoarding of food for later digestion is really a type of farming carried out by the foraminifera. It is easy to see how symbiotic relationships could have evolved stepwise. More digestion-resistant food would have a greater chance of survival within the host. Some evidence for this hypothesis was obtained from infection experiments. *Rosalina leei* ate large numbers of *C. hedleyi* but they resisted digestion (Lee and Zucker, 1969). Large numbers were still packed inside the last chamber of the foraminifera a week after feeding. Although there is no experimental evidence it is conceivable that the attraction between *Dunaliella* (a genus closely related to *Chlamydomonas*) and foraminifera may also have been one of the factors initiating symbiosis in *Archaias*. It is difficult to generalize about the morphological adaptation or preadaptation which *Archaias* and many soritids seem to have for symbiosis because we know so little about the breadth of symbiosis in the group. Disc-shaped *Archaias* with its packed symbionts clustered under thin window-like chamberlets remind one of plants in a greenhouse. The morphology of the soritid and nummulitid groups seem ideal for the establishment and maintenance of symbiotic algae.

Chlamydomonas hedleyi is well adapted as a symbiont since it grows best with urea as a nitrogen source. This could be a waste product of the host and might be available in concentrations which would favor the growth of the symbiont. It is not unique in this respect since many free living forms can also use it (Carpenter *et al.*, 1972, Guillard, 1963, Hayward, 1965, Lui and Roels, 1972, Pintner and Provasoli, 1963, Ryther, 1954, Syrett, 1962). The amino acids and thiamin which stimulated the growth of *C. hedleyi* might be available in excess quantities from food digested by the host. Thiamin stimulation might be construed as evidence that other organisms might also be in association with the host since this vitamin is not too stable in sea water (Gold, 1968, Gold *et al.*, 1966). During the process of isolation of *C. hedleyi*, 2 species of bacteria were consistently associated with fragments of the host shell. It is possible that these bacteria might provide thiamin for growth of the symbionts.

E. Some Current Problems

Too little is known about symbiosis in benthic foraminifera to make exacting comparisons about the phenomenon in various species. Present evidence suggests that the diatom zooxanthellae in *Heterostegina depressa* are much more important to their hosts' carbon budget than *C. hedleyi* is to *Archaias*. Uptake of $H^{14}CO_3^-$ by the *Archaias* symbiont system was much more rapid in fed than in starved organisms (Lee and Zucker, 1969). However, increases in light enhanced calcification of the host. In contrast starved *H. depressa* grew well in optimal light (300 lux) (Rottger and Berger, 1972). Fed organisms in the dark did not grow at all but both starved and fed animals survived for 4 months in the dark (Dietz-Elbrachter, 1971).

Although we know more about the symbiotic relationship in *Archaias* than in any other species of foraminifera, many interesting questions still remain. Among them would be questions on the overall carbon budget of the host, the range of hosts which can harbor *Chlamydomonas hedleyi*, the range of algae which can establish a symbiotic relationship with *Archaias*. What happens to the symbionts and the host during starvation? Are the symbionts ever released free-living from the host? Are juvenile *Archaias* released from their parents with symbionts as in *Heterostegina depressa* or are they infected later? Similar questions about gnotobiotic cultures of other species of foraminifera and their symbionts are also germane.

During the course of field studies on the semitropical foraminifera in Biscayne Bay, Bimini, and Bermuda the author has often observed symbionts in soritids from the region. They have also been reported in foraminifera from Tortugas and Cuba (Doyle and Doyle, 1940, Cushman, 1922, 1955). Superficial examination of the symbionts expressed from *Peneroplis*, *Spirolina*, some specimens of *Archaias*, *Sorites*, *Orbitolites* and an occasional specimen of *Marginopora* suggests that they harbor dinoflagellate zooxanthellae very similar to *Symbiodinium microadriaticum*. Circumstantial evidence suggests that they may in fact be *S. microadriaticum* because *Cassiopea*, a host for *Symbiodinium*, is very abundant in the same waters. It is the hope of this author someday to collect, culture, and isolate these soritid foraminifera and their symbionts and to study, as comparatively as possible, the role of the symbionts in the overall carbon budget of their hosts.

V. Concluding Remarks

It is clear from the foregoing account that we have touched on only a few aspects of the niches of some salt marsh and estuarine foraminifera. The negative interactions (predation, parasitism and antagonism) have not been discussed at all. Yet if we can generalize from the interpretations of Paine (1966) in his review of food web complexity and species diversity, predation may also be an important determinant for foraminiferan diversity. Paine showed a direct relationship between local species diversity patterns on rocky intertidal habitats and predation. When he removed the top carnivore, *Pisaster*, (a starfish), from an experimental plot, species diversity sharply declined. *Pisaster* is capable of removing 20–60 barnacles at a time and thus enhances the ability of other species to colonize the area. Other smaller predators, species of *Thais* (carnivorous snails), increased in numbers 10 to 20 fold in the absence of *Pisaster* but did not prevent the eventual monopolizing of the shore by *Mytilus californianus*. Quite a number of animals have been reported to eat foraminifera. Lipps and Valentine (1970) have drawn together in their recent review quite a number of diverse observations on these animals which include gastropods (both prosobranch and opisthobranchs) scaphopods, pteropods, various crustacean and echinoderm groups and burrowing worms. Rigorous quantitative assessment of this phenomenon is needed if we are ever to understand the role which foraminifera play in the trophic dynamics, energetics, and mineral cycling of the communities of which they are an important part. There could be some paleoecological significance to this aspect of foraminiferal ecology if it can be demonstrated that there is selective destruction of species by predators, particularly deposit feeders.

Many other questions about the distribution of foraminifera and their niches are still unanswered. Patchiness is an ecological strategy which diversifies the habitat and reduces competition. The dietary habits of salt marsh foraminifera and the characteristics of the algal assemblage suggest that the 2 groups should be linked both temporally and spatially. Critical experimental demonstrations are still needed on this point. How are foraminiferal life cycles adapted to the dynamic changes in the community? Do species of foraminifera grow and reproduce in response to particular algal blooms? Such specialists would maximize energy yield by decreasing search time per food organism eaten and would temporarily have very high ecological efficiencies but might pay for this later in energy of maintenance between blooms of food. What triggers reproduction, algal blooms or anticipation of food for offspring? If a foraminiferal species has a long generation time will its offspring be widely adapted to feed on the new organisms available in the community at the time of their release (feeding generalist) or will the offspring grow slowly and survive until the next bloom (possibly half a year or more later; feeding specialist)? Probably both types of feeding strategies have evolved among the foraminifera.

The role of substrate in developing communities which will become suitable habitats for foraminifera is a premier intellectual question, one step further removed from the above, but within our present experimental technology. If it can be demonstrated that foraminifera are closely linked to habitat by dietary restrictions then comparative studies of the microbial and algal assemblages of foraminiferal blooms and foraminifera-barren communities should prove fruitful. More refined studies of those relationships discussed earlier in which the distribution of particular species has been linked to particular substrates (i.e., *E.*

incertum and *T. inflata* and sediment median grain size) would be of great interest. Such studies would be pioneering, judging from the recent review by Round (1971) since very little information is available on the comparative population structure or small scale patchiness of relevant aufwuchs or psammolittoral communities.

Studies of intraspecific and interspecific competition among foraminifera and between foraminifera and other microherbivores have just barely scratched the surface of this important aspect of niche. Presumably intraspecific competition is an important aspect of life cycles leading to, or promoting, dispersion of offspring. One could expect intraspecific competition to be high in species with apogamic life cycles (i.e., *Allogromia laticollaris*, Table 19) and lower in those with more obligate sexual cycles or in the gamontic generation. *A priori* adaptive strategies evolve toward maximizing total net energy yield. The energy expense of foraging for food may be reduced where animals are not grouped too closely but then the number of contacts between mates becomes so low some other mechanism(s), resulting in the expenditure of energy, must be operative for attracting gamonts or gametes.

Interspecific competition depends on many variables which have not yet been seriously studied by foraminiferal specialists. The simplest form of competition for food exists when 2 or more species of herbivores feed simultaneously on the same food species. For competition to take place there must be less food available than the 2 species require. In aufwuchs communities there may be a number of species of herbivores or omnivores which may attack the algal species temporarily in abundant supply. The extent to which the same species can also exploit other sources of food would seem to be of great adaptive value as hedge against competitive pressure since they could shift to their alternate foods. Most herbivores in the aufwuchs community probably feed on several species of prey even though each has "preferred" food so some overlap is expected. Quantification, or even precise definition of competitive interactions between foraminifera and other microherbivores will not be easy but should be attempted. Some aspects of the types of competitive adaptations available to foraminifera are inherent in their relatively large size, long life spans and long generation times when compared to many other types of marine protozoa and some micrometazoa. (For an up to date review of the feeding strategy aspect of competition, see Schoener, 1971).

As our views of foraminiferal ecology have gradually evolved into more quantitative analyses and detailed approaches to explaining the bases of species diversity in foraminifera, the techniques for doing so have become more sophsticated and correspondingly more time consuming. It is with some trepidation and some hope that we look forward to future progress in this field.

VI. Acknowledgments

The author gratefully acknowledges the meaningful discussions he has held on the subject with his colleagues Drs. Norman Newell, Roger Batten and Niles Eldridge at The American Museum of Natural History, and Drs. John H. Tietjen and William A. Muller at The City College.

VII. Glossary

Agnotobiotic—pertaining to a culture or system, usually maintained under septic conditions, in which all the organisms present are not characterized.

Aseptic—free or freed from contaminating microorganisms. Usually used in the context of techniques or procedures which are performed with sterile instruments in a germ-free environment.

Aufwuchs—a community of microorganisms, small animals, and small plants living attached to, crawling or creeping on, or otherwise associated with the stems and leaves of rooted aquatic plants or other surfaces projecting above the benthos.

Auxotrophic—a characterization of the nutritional requirements of an organism; considered by an expert arbitrarily to be more complex than the minimum for its type.

Axenic—a subclass of gnotobiotic. A pure culture (free from all other organisms).

Biotic potential—the overall population growth rate under unlimited environmental conditions. The difference between the biotic potential and the reproductive rate measured in the laboratory or in the field is taken as a measure of environmental resistance.

Carrying capacity—a density dependent relationship. The maximum population of a species which will be supported by a stipulated environment without the population or the environment undergoing some form of deterioration which will result eventually in diminished population size.

Clone—a population of organisms which are the asexual progeny of a single individual.

Cytotomy—a form a multiple fission (asexual reproduction) which has some characteristics of budding, binary fission, and schizogony. In cytotomy, the parent organism sequentially pinches off rather large offspring and in the end is itself reduced to the size of some of its daughters.

Diurnal rhythm—a rhythm which occurs daily during the day. Other rhythms: diel—events which recur at 24 h intervals; circadian—events which recur $\sim$ every 24 h, regulated by biological clocks which couple environmental and physiological events.

Epibenthic—organisms which live on the surface of the bottom sediments.

Ecological growth efficiency—$(E_e = P/I)$. One of the various measures of the efficiency of food web transformations. It equals the production of new biomass (P) divided by the total energy intake (I).

Environmental grain—A quality of a habitat and the populations within it. In a coarse grained environment the organisms move infrequently from one habitat to another whereas in a fine grained environment members move readily from habitat to habitat.

Epiphyte—organisms, usually microorganisms, small plants and animals, which live on the surface of larger plants.

Generation time—the time it takes for any population of organisms to double in number.

Gnotobiotic—a culture or system in which all the organisms present are known. Opposite: agnotobiotic.

Heterotrophic—pertaining to a nutritional type; requiring organic molecules as energy sources and for growth.

Intrinsic rate of natural increase (r)—a measurement of population growth in an unlimited environment under specified physical conditions. It is a measure at any instant of the specific birth rate minus the death rate. At maximum (r_{max}) it is called biotic potential.

Logarithmic growth phase—a characteristic phase of a culture of microorganisms in which the numbers of cells are increasing at an increasing rate (logarithmically).

Macrophyte— a large aquatic plant; commonly used synonym for seaweeds.

Manometric technique—a technique which measures changes in gas pressure. Commonly used to measure the respiration of small organisms.

Medium spectrum index (M.S.I.)—a measure of the ability of a medium to support unselectively the growth of large numbers of species. A medium which has a $MSI < 1$ is more selective than sea water from the same site. Larger numbers of fewer species would grow on such a medium.

Meiofauna and meioflora—operationally defined terms. Small animals and plants characterized by their ability to pass through all but the smallest sieves ($\sim$0·075 mm) used to study sediments. Their size range is generally $\sim$50 − 500 μm.

Monoxenic—pertaining to a gnotobiotic culture which contains 2 species of organisms, a species being studied and a cultural associate (usually food).

Niche, habitat (older concept)—a site or habitat supplying those factors characteristically necessary for the successful existence of an organism or species.

Niche, fundamental—an abstract concept developed by G. E. Hutchinson in which niche is formalized as a maximally inhabitable, 3 dimensional geometric figure (hypervolume) constructed through consideration of orderable environmental variables within which the species could survive indefinitely without biotic constraints.

Niche, realized—a smaller hypervolume occupied under biotic constraints. A set of elements within the environment which is occupied by a species under "usual" environmental conditions and in the presence of normal competitors, parasites and predators.

Niche breadth—a comparative term. A measure of the ability of an organism to occupy a variety of habitats and to tolerate a range of biotic and abiotic environmental conditions.

Operational habitat—an identifiable unit which is lived in, occupied by, or foraged by the organism of reference.

Psammolittoral community (= interstitial flora and fauna) referring to an ecological community consisting of typically minute plants and animals which live in the water filling the interstices of sand or other sediments in a littoral zone.

Species diversity index—a measure of the complexity of a community, calculated in various ways, but generally a ratio between the number of species and either total numbers of organisms, biomass, or productivity. Species diversity tends to be low in physically controlled ecosystems and high in biologically controlled ones.

$$d = \frac{S-1}{\log N} \quad d = \frac{S}{\sqrt{N}}$$

S = number of species
N = number of individuals.

Synergistic—an interaction of 2 or more biotic or abiotic factors on an organism, whose total effect is greater than the effect of both factors operating alone.

Synxenic—A gnotobiotic culture in which the species of reference is being cultured in the presence of one or more additional species of organisms (usually food). Types of synxenic culture are monoxenic, dixenic, etc.

Zoochlorella—A general term for any green symbiotic alga. They belong to Chlorophyceae and Prasinophyceae.

Zooxanthella—a general term for any olive, brownish-red or golden brown symbiotic alga. They belong to the Pyrophyceae Cryptophyceae and rarely Chrysophyceae.

VIII. References

Aleem, A. A. (1950). *J. Ecol.* **38,** 75–106.

Angell, R. W. (1967). *Cushman Found. foramin. Res.* **18,** 176–177.

Antia, N. J., McAllister, C. D., Parson, T. R., Stephens, K. and Strickland, J. D. H. (1963). *Limnol. Oceanogr.* **8,** 166–183.

Arnold, Z. (1954a). *Cushman Found. foramin. Res.* **5,** 4–13.

Arnold, Z. (1954b). *J. Paleontol.* **28,** 78–101.

Arnold, Z. (1966). *Micropaleontol.* **12,** 109–118.

Arnold, Z. (1967). *Vie milieu* (ser. A) **18,** 36–45.

Belar, K. (1923). *Arch. Protistenk.* **46,** 1–96.

Belar, K. (1924). *Arch. Protistenk.* **48,** 371–435.

Boltovskoy, E. (1964). *J. Cons. Perm. Int. Explor. Mer.* **39,** 136–145.

Boltovskoy, E. and Lena, H. (1969). *Rev. Micropaleontol.* **12,** 177–185.

Bradshaw, J. S. (1957). *J. Paleontol.* **31,** 1138–1147.

Bradshaw, J. S. (1961). *Cushman Found. foramin. Res.* **12,** 87–106.

Bradshaw, J. S. (1968). *Limnol. Oceanogr.* **13,** 26–38.

Brooks, A. L. (1967). *Limnol. Oceanogr.* **12,** 667–684.

Bunt, J. S. (1969). *J. Phycol.* **5,** 37–42.

Burkholder, P. R. (1963). *In* "Symposium on Marine Microbiology (C. H. Oppenheimer, ed.) pp. 135–150. C. C. Thomas, Springfield, Illinois.

Butschli, O. (1886). *Morph. Johrb.* **11,** 78–101.

Buzas, M. A. (1965). *Smithsonian Misc. Coll.* **15,** 8.

Buzas, M. A. (1970). *Ecology* **51,** 874–879.

Carlucci, A. F. and Bowes, P. M. (1970a). *J. Phycol.* **6,** 351–357.

Carlucci, A. F. and Bowes, P. M. (1970b). *J. Phycol.* **6,** 393–400.

Carpenter, E. J., Remsen, C. C. and Watson, S. W. (1972). *Limnol. Oceanog.* **17,** 265–269.

Cushman, J. A. (1922). *Papers Tortugas Lab.* **17,** 1–85.

Cushman, J. A. (1955). "Foraminifera, Their Classification and Economic Use". Harvard University Press, Cambridge.

Dietz-Elbrachter, G. (1971). *Forsch-Ergebnisse C.* **6,** 41–47.

Doyle, W. L. and Doyle, M. M. (1940). *Papers Tortugas Lab.* **32,** 127–142.

Ellison, R. L. and Nichols, M. M. (1970). *Contrib. Cushman Found. foramin. Res.* **21,** 1–17.

Fenchel, R. (1968). *Ophelia* **5,** 73–121.

Fenchel, R. (1972). *Deuts. Zool. Gesel.* **65,** 14–22.

Fenchel, T. and Jannson, B. O. (1966). *Ophelia* **3,** 161–177.

Fogg, G. E. (1966). *In* "Advances in Marine Biology" (H. Barnes ed.). Vol. 4, 195–212. Academic Press, London and New York.

Fogg, G. E. (1971). *Arch. Hydrobiol. Beih. Ergebn. Limnol.* **5,** 1–25.

Føyn, B. (1936). *Bergens Museum Aabok.* **2,** 1–16.

Freudenthal, H. D. (1962). *J. Protozool.* **9,** 45–52.

Gold, K. (1968). *Limnol. Oceanog.* **13,** 185–188.

Gold, K., Roels, O. A. and Banks, H. (1966). *Limnol. Oceanogr.* **11,** 410–413.

Guillard, R. R. L. (1963). *In* "Symposium on Marine Biology" (C. H. Oppenheimer, ed.) pp. 93–104. C. C. Thomas, Springfield, Illinois.

Hargrave, B. T. (1970). *Limnol. Oceanog.* **15,** 21–30.

Hayward, J. (1965). *Physiol. Plant.* **18,** 201–207.

Hedley, R. H. (1964). *Int. Rev. Gen. Exptl. Zool.* **1,** 1–45.

Hellebust, J. A. (1965). *Limnol. Oceanogr.* **10,** 192–206.

Hellebust, J. A. (1970). *In* "Symposium on Organic Substances in Natural Waters" (D. W. Hood, ed.) pp. 225–265. *Inst. Mar. Sci. Occasional Publ. No. 1,* University of Alaska, Fairbanks.

Hellebust, J. A. (1971). *J. Phycol.* **7,** 1–4.

Hellebust, J. A. and Guillard, R. R. L. (1967). *J. Phycol.* **3,** 132–136.

Hellebust, J. A. and Lewin, J. (1972). *Can. J. Microbiol.* **18,** 225–233.

Horne, A. J., Fogg, G. E. and Eagle, D. J. (1969). *J. Mar. Biol. Assoc. U. K.* **49,** 393–405.

Hovasse, R. (1923). *Bull. Soc. Zool. France* **48,** 247–254.

Huntsman, S. A. (1972). *J. Phycol.* **8,** 59–63.

Jepps, M. W. (1926). *Q. Jl. microrc. Sci.* **70,** 701–720.

Jepps, M. W. (1942). *J. Mar. Biol. Assoc. U. K.* **25,** 608–666.

Jepps, M. W. (1956). "The Protozoa, Sarcodina". Oliver and Boyd, London.

Johannes, R. E. and Webb, K. L. (1965). *Science* **150,** 76–77.

Kevin, M. J., Hall, W. T., McLaughlin, J. J. A. and Zahl, P. A. (1969). *J. Phycol.* **5,** 341–350.

LeCalvez, J. (1938). *Arch. Zool. Exp. Gen.* **80,** 163–333.

Lee, J. J., Crockett, L. J., Hagen, J. and Stone, R. J. *The taxonomic identity and physiological ecology of* Chlamydomonas hedleyi *(the algal flagellate symbiont from the foraminifera* Archaias angulatus). Ms submitted for publ.

Lee, J. J., Freudenthal, H. D., Muller, W. A., Kossoy, V., Pierce, S. and Grossman, R. (1963). *Micropaleontology* **9**, 449–466.

Lee, J. J., Freudenthal, H. D., Kossoy, V. and Be, A. (1965). *J. Protozool.* **12**, 531–542.

Lee, J. J. and McEnery, M. E. (1970). *J. Protozool.* **17**, 184–195.

Lee, J. J., McEnery, M. E., Kennedy, E. M. and Rubin, H. (1973). *In* "Modern Methods in the Study of Microbial Ecology". *Bull. Ecol. Res. Comm, Stockholm.* **17**, 387–397.

Lee, J. J., McEnery, M., Pierce, S., Freudenthal and Muller, W. A. (1966). *J. Protozool.* **13**, 659–670.

Lee, J. J., McEnery, M. E. and Rubin, H. (1969b). *J. Protozool.* **16**, 377–395.

Lee, J. J. and Muller, W. A. (1973). *In* "Conference on Culture of Invertebrate Animals", Middle Atlantic Science Council, Inc., Greenport, N.Y., Oct. 5–7, 1972. In press.

Lee, J. J. and Muller, W. A. (1973). *Amer. Zool.* **13**, 215–223.

Lee, J. J., Muller, W. A., Stone, R. J., McEnery, M. and Zucker, W. (1969a). *Mar. Biol.* **4**, 44–61.

Lee, J. J. and Pierce, S. (1963). *J. Protozool.* **4**, 404–411.

Lee, J. J., Pierce, S., Tentchoff, M. and McLaughlin, J. J. A. (1961). *Micropaleont.* **7**, 461–466.

Lee, J. J., Tietjen, J. H., Stone, R. J., Muller, W. A., Rullman, J. and McEnery, M. (1970). *Helgolander wiss. Meeres.* **20**, 136–156.

Lee, J. J. and Zucker, W. (1969). *J. Protozool.* **16**, 71–81.

Lewin, J. and Hellebust, J. A. (1970). *Can. J. Microbiol.* **16**, 1123–1129.

Lewin, J. and Lewin, R. A. (1960). *Can. J. Microbiol.* **6**, 127–134.

Lewin, J. and Lewin, R. A. (1967). *J. Gen. Microbiol.* **46**, 361–367.

Lipps, J. J. and Valentine, J. W. (1970). *Lethaia* **3**, 279–286.

Lister, J. J. (1895). *Phil. Trans. R. Soc.* **186**, 401–454.

Lui, N. S. T. and Roels, O. A. (1972). *J. Phycol.* **8**, 259–263.

Lutze, G. F. (1968). *Meyniana* **18**, 13–30.

Lynts, G. W. (1966). *Limnol. Oceanogr.* **11**, 562–566.

Matera, N. J. and Lee, J. J. (1972). *Mar. Biol.* **14**, 89–103.

McEnery, M. E. and Lee, J. J. (1970). *J. Protozool.* **17**, 184–195.

McIntire, C. D. and Overton, W. S. (1971). *Ecology* **52**, 758–777.

Muller, W. A. (1972). "Graphic representation of niche width and its application to salt marsh littoral foraminifera". Ph.D. thesis, City University of New York.

Muller, W. A. and Lee, J. J. (1969). *J. Protozool.* **16**, 471–478.

Murray, J. W. (1963). *J. Mar. Biol. Assoc. U. K.* **43**, 621–642.

Myers, E. H. (1935). *Trans. Amer. Micros. Soc.* **54**, 264–267.

Myers, E. H. (1936). *J. Roy. Soc.* **61**, 120–146.

Myers, E. H. (1937). *In* "Culture Methods for Invertebrate Animals" (P. S. Galstoff, ed.) pp. 93–96. Comstock, Ithaca, New York.

Myers, E. H. (1942a). *J. Paleontol.* **16**, 397–398.

Myers, E. H. (1942b). *Proc. Am. phil. Soc.* **85**, 325–342.

Myers, E. H. (1943). *Proc. Am. Phil. Soc.* **86**, 439–458.

Paine, R. T. (1966). *Am. Nat.* **100**, 65–75.

Phleger, F. B. and Bradshaw, J. S. (1966). *Science* **154**, 1551–1553.

Phleger, F. B. and Lankford, R. R. (1957). *Contrib. Cushman Found. foramin. Res.* **8,** 93–105.

Pintner, I. and Provasoli, L. (1963). *In* "Symposium on Marine Microbiology". (C. H. Oppenheimer, ed.) pp. 114–121. C. C. Thomas, Springfield, Illinois.

Rhumbler, L. (1911–1913). Humboldt-Stiftung Plankton Expedition. *Ergebn.* **3,** 1–476.

Ross, C. A. (1972). *J. Protozool.* **19,** 181–192.

Rosswall, T. (ed.) (1973). "Modern Methods in the Study of Microbial Ecology". *Bull. Ecol. Res. Comm. (Stockholm)* **17.**

Rottger, R. (1972a). *Mar. Biol.* **15,** 150–159.

Rottger, R. (1972b). *Mar. Biol.* **17,** 228–242.

Rottger, R. and Berger, W. A. (1972). *Mar. Biol.* **15,** 84–89.

Round, F. E. (1971). *Oceanogr. Mar. Biol. Ann. Rev.* **9,** 83–139.

Ryther, J. H. (1954). *Biol. Bull.* **106,** 198–209.

Saks, N. M. and Lee, J. J. (1972). Third Nat. Sympos. Radioecology, in press.

Saks, N. M. and Lee, J. J. (1973). *In* "Radioactive Contamination of the Marine Environment" Symposium, IAEA-SM-158/35. pp. 565–571.

Saks, N. M., Lee, J. J., Muller, W. A. and Tietjen, J. H. (1973). Thermal Ecology Symposium, in press.

Saks, N. M., Lee, J. J. and Stone, R. J. (1973). *Carbon budget of salt marsh epiphytic algae.* ms in preparation.

Samuels, S., Shaw, W. M. and Fogg, G. E. (1971). *J. Mar. Biol. Assoc. U. K.* **51,** 793 798.

Sandon, H. (1932). *Pub. Fac. Sci. Egypt Univ.* **1,** 1–187.

Schoener, T. W. (1971). *Ann. Rev. Ecol. System.* **2,** 369–404.

Shifflet, E. (1961). *Micropaleontology* **7,** 45–54.

Smith, D. L., Muscatine, L. and Lewis, D. (1969). *Biol. Rev.* **44,** 17–90.

Syrett, P. J. (1962). *In* "Physiology and Biochemistry of Algae" (R. A. Lewin, ed.). pp. 171–188. Academic Press, London and New York.

Taylor, D. (1968). *J. Mar. Biol. Assoc. U. K.* **48,** 349–366.

Tietjen, J. H. (1969). *Oecologia* **2,** 251–291.

Tietjen, J. H. and Lee, J. J. (1972). *Oecologia* **10,** 167–176.

Tietjen, J. H., Lee, J. J., Rullman, J., Greengart, A. and Trompeter, J. (1970). *Limnol. Oceanogr.* **15,** 535–543.

Vandermeer, J. H. (1972). *Ann. Rev. Ecol. System.* **3,** 107–133.

Winter, F. W. (1907). *Arch. Protistenk.* **10,** 1–113.

Zucker, W. H. (1973). "Fine structure of planktonic foraminifera and their endosymbiotic algae". Ph.D. Thesis, City University of New York.

Note added in proof: During 1973 my colleagues and I studied the symbionts in *Sorites* and *Archaias* from Key Largo Sound, Florida. Primary production and respiration of *Sorites* and *Archaias*-symbiont systems were measured *in situ*. *Sorites* symbionts fixed more net carbon (3.1×10^{-4}M carbon/mg dry organic weight/h) than did *Archaias* symbionts (0.5×10^{-5}M carbon/mg dry organic weight/h). Preliminary light and electron microscopic studies suggest that *Sorites* harbors at least two kinds of symbionts. One *Sorites* symbiont is a dinoflagellate similar to, but perhaps not identical to, *Symbiodinium microadriaticum*.

Great Names in Micropalaeontology

1. Alcide d'Orbigny

by

Y. LE CALVEZ

*École Pratique des Hautes Études, Laboratoire de Micropaléontologie,
8 rue de Buffon, Paris Ve, France*

Attempting a new account of the life and work of Alcide d'Orbigny after the outstanding biography published in 1917 by Heron-Allen, is no easy task. Yet it is not possible to deal with the "Great Names in Micro-palaeontology" without beginning with the man who was truly the founder of the subject; the scholar to whom we owe so many of the fundamental facts concerning foraminifera. Although d'Orbigny accomplished much in other fields, notably stratigraphy and macro-palaeontology, comment is here reserved exclusively for his micro-palaeontological publications.

Alcide d'Orbigny was born in Couëron (Charente-Maritime) on September 6th, 1802. His family, records of which can be traced back to the 15th century when barons of that name were in the service of King Louis XI, later emigrated to San Domingo in the Caribbean and was almost totally exterminated during the revolt of the black slaves. Of the eighteen d'Orbigny children, only two sons, in France to complete their education, escaped the massacre. One of them, Charles-Marie d'Orbigny married Marie-Ann Pipat in 1799, this union producing five children of whom one was Alcide d'Orbigny. A naval surgeon by profession, Charles-Marie eventually settled at Esnandes on the coast of Vendée, where he practised medicine and was able to pursue his interest in marine life as an amateur naturalist. Indeed, he and his friend Fleuriau de Bellevue assembled a collection of shells of the region and founded our first regional museum—that of La Rochelle.

Alcide d'Orbigny and his brother Charles were thus initiated into Natural History at an early age. They searched for and examined shells, and were enabled to portray them owing to their precocious talents for drawing. Alcide was only 11 years old when he began to examine fora-minifera found in the local beach sands, this activity soon being followed by studies of material from Rimini and Corsica. Until this time all microscopic shells were classed with the "Céphalopodes polythalames" of Lamarck, and no one had tried to separate them. D'Orbigny undertook this great task. In April 1825 he went to Paris, examined the collections of Defrance and Lamarck, and some months later (1826) published his "Tableau méthodique de la classe des Céphalopodes" in which he distinguished the microscopic species from the other cephalopods, and called them Foraminifera. In a report to the Academy of Sciences on 12th December 1825, Geoffroy Saint-Hilaire et Latraille was able to write on this subject "nous voilà arrivés au 3 ème ordre, celui des Foraminifères, le plus compliqué et le plus embarrassant de tous, vu la quantité presque innombrable d'espèces dont il se compose et leur extrême petitesse, ordre qui est une création de Monsieur d'Orbigny".

This work attracted much attention at the time, and led to its author being charged by the Museum d'Histoire Naturelle with a mission to South America. D'Orbigny left in June 1826 and until March 1834 he travelled the American continent from "the cold and arid regions of Patagonia to the torrid zone, from sea level to the highest plateaux of the Andes, from the shores of two oceans to the centre of the continent; visiting in turn, Brazil, the Pampas of Buenos Aires, the borders of Paraguay, and the republics of Chile, Bolivia and Peru" as he himself wrote in his detailed report of 1850.

This long voyage enabled him to amass a great quantity of material, the study of which occupied the years following his return, the results being expressed in the publication of nine volumes and about 500 plates. Amongst these were three of particular concern to us: "Les Foraminifères de l'Amérique méridionale", "les Foraminifères de l'île de Cuba et des Antilles" and "les Foraminifères des îles Canaries", all of which were published in 1839.

In his study of the Foraminifera of Cuba, d'Orbigny gave a detailed historical account of the work carried out on these organisms between 1731 and 1839 before describing his own observations made during twenty years of research. He stressed the coloration, "yellow, fawn, russet, red-violet or bluish", of the cytoplasm, and the length of the pseudopodia, which he called "filaments" attaining up to five or six times the diameter of the body, thus indicating that his observations, far from being restricted to the shell, also extended to the living animal. He then proposed a classification of the Foraminifera based on their mode of growth and showing the relationships of the orders.

But while preparing these works, d'Orbigny also concerned himself with other matters. He turned to Palaeontology and interested himself in the foraminifera of the Chalk, a study which led to the production of a large memoir (1839) in which fifty four species were described and figured, all but three or four being new. These observations, moreover, led him to think that during the Cretaceous the Paris Basin was invaded by a warm sea lacking strong currents, and to conclude that species diversity increased with the passage of time, i.e., towards the present day.

In 1846 his study of the fossil foraminifera of the Vienna Basin appeared—a work intended to be a complete exposition of the Foraminifera. Everything was included: systematics, with the description and illustration of two hundred and twenty eight species; geological distribution (Palaeozoic to Recent); the proportinate increase in the number of genera and species as the present epoch is approached; the characterization of warm regions by some forms and of cold regions by others, each species being more or less restricted to a particular area; the comparison of Recent and fossil genera. In other words, most of the problems posed by the study of foraminifera were already foreseen if not resolved by d'Orbigny, who wrote in 1846, "ils peuvent servir à déterminer sûrement l'âge d'un terrain géologique" thus anticipating the preeminent place which oil geologists would reserve for these microfossils one hundred years later.

Finally, d'Orbigny worked for more than 14 years on the "Prodrome de Paléontologie stratigraphique des animaux Mollusques rayonnés". This immense work, to which he devoted so much time yet never actually finished, appeared in 1850–52 in the form of a stratigraphical list of genera and species. The illustrations, in the form of plates, remained unavailable until their publication (as outline drawings only) by Fornasini between 1897 and 1908.

Having become the acknowledged authority in the field of Palaeontology, d'Orbigny sought a post as professor, but he was opposed by many zoologists who did not accept his discoveries, and by geologists who scarcely appreciated his views on stratigraphical terminology. At last, by a decree of 1853, he obtained a Chair of Palaeontology in the Museum d'Histoire Naturelle. However, he hardly profited by it, for he died on June 30th, 1857, at the early age of 55.

The study of foraminifera represents but a small part of d'Orbigny's immense output of admirable work, and I have given here only an outline of the publications that concern us. One is overwhelmed by the number of genera and species which he described; by the drawings, the originals of which are preserved in the Palaeontological Institute of the Museum d'Histoire Naturelle, and which testify to the ability of the man who executed them as well as to the quality of his observations; by the models, sculpted with his own hands and carefully reproducing the shells of these microscopic animals to an enlargement of from 40–200 times.

D'Orbigny has been reproached for his anti-evolutionary ideas: in his view, microfaunas disappeared at the end of each stage, making way for new associations composed of different species. Moreover, he did not believe that species enjoyed a wide geographical distribution, and thus gave new names to similar individuals found in the Mediterranean and

Antilles. However, in certain cases he realized that he was mistaken. For example, *Planorbulina mediterranensis* was renamed *vulgaris* when he realized that this foraminifer, first discovered in the Mediterranean, also occurred off Cuba.

D'Orbigny's conceptions are explained by the fact that he lived at the dawn of micropalaeontology—at a time when the shell was all that was known of the animal. How was it possible to speak of evolution in animals which lacked organs, and when the fossiliferous strata studied were of such different ages that the connecting links between the genera and species were not visible?

Today, d'Orbigny still remains the first and most famous of all micropalaeontologists; the first to recognize the importance of foraminifera and to foresee the place they would one day occupy in stratigraphical geology.

Systematic Index

(Sarcodina only)

A

Abathomphalidae, 35
Abathomphalus, 35
Aboudaragina, 33
Abrardia, 14
Acarininae, 37
Acarinininae, 37
Accordiella, 14
Acervulinidae, 41, 48
Aciculella, 9
Aciculina, 9
Actinosiphon, 38
Adelungia, 8
Aeolostreptis, 32
Afrobolivina afra, 59
Agathammina, 18
  pusilla, 99
Aktinocyclina, 38
Aktinorbitoides, 38
Alabaminidae, 33, 39, 41, 49
Alfredininae, 32
Alliatina, 5
Alliatininae, 30, 34
Allogromia, 160, 224, 225, 236
  laticollaris, 160, 191, 213, 216, 222, 224—229, 233, 234, 237—240, 242, 244, 254
  priori, 254
Allogromiidae, 7, 41, 42
Allogromiina, 6, 7, 42
Allogromiinae, 7
Allomorphinella, 39
Almaeninae, 49
Alveocyclammina, 14
Alveolina, 98, 99
  munieri, 99
  porrecta, 99
  stipes, 99
  tenuis, 99

Alveolina, *(contd.)*—
  viola, 99
Alveolinidae, 18, 41, 46, 98
Alveolophragmium, 14
  A. (Reticulophragmium), 14
Alveosepta, 14
Alveovalvulinella, 15
Ambitropus, 33
Ammobaculites dilatatus, 213
Ammodiscacea, 8, 40—42
Ammodiscidae, 9, 11, 40, 41, 43
Ammodiscinae, 11, 40, 43
Ammodiscus, 11
Ammoglobigerina, 12
Ammonia, 2, 20, 22, 23, 37
  beccarii, 213—216, 222, 225, 226, 233, 234, 236, 238, 243, 244
  —var. *tepida*, 226, 232
Ammotium salsum, 213, 214
Ammovertillina, 11
Ammovolummina, 11
Ammovolummininae, 8, 11, 43
Amoeba, 209
Amphisteginidae, 41, 48
Amphitrematinae, 7
Andersenia, 12
Angulodiscorbis, 29, 33
Annulopatellinidae, 38, 41, 49
Anomalina, 233
Anomalinacea, 4, 22, 39
Anomalinidae, 33, 39, 41, 49
Anomalininae, 49
Aoujgalia, 15
Archaediscidae, 16, 29, 41, 45
Archaediscinae, 45
Archaeglobigerina, 35, 37
Archaias, 17, 246, 251, 252
  angulatus, 246
Archaiasinae, 46
Arenbulimina, 13
Arenoturrispirillina, 11
Armorella, 8, 11

Asanospira, 11
Asanospiridae, 9, 11, 12
Aschemocellidae, 9, 11
Aschemocellinae, 43
Aschemonella, 11
Aschemonellinae, 11, 43
Astacolinae, 27
Astacolus, 25—27
Asterellina, 32
Asterigerina, 23
Asterigerinatinae, 32
Asterigerinidae, 32, 33, 41, 48
Asterobis, 38
Asterocyclina, 38
Asterophragmina, 38
Astrammina, 8
Astrorhiza limicola, 7, 160
Astrorhizidae, 9, 10, 40—42
Astrorhizidea, 9
Astrorhizinae, 10, 42
Asterorotalia, 37, 106
   A. (Clavatorella), 37
Asymmetrina, 30, 39
Asymmetrinidae, 30, 40—42, 47
Ataxogyroidina, 13
Ataxoorbignyna, 13
Ataxophragmiidae, 9, 12, 13, 15, 41, 44
Ataxophragmiinae, 12, 13, 44
Ataxophragmium, 13
Athecocyclina, 38
Aulotortus, 29

B

Baggininae, 33, 47
Barbourinella, 12
Barkerina, 13, 19
Barkerinidae, 13, 18, 19, 41, 43
Bathysiphon, 11
Bathysiphoninae, 11
Beella, 37
Biapertorbis, 33
Biconcava, 13
Bilamellidea, 4
Biplanata, 13
Biseriamminidae, 41, 44
Biticinella, 37
Biumbella, 15
Boderia turneri, 7, 160
Bogdanowiczia, 10
Bolivina, 31, 72, 102, 222, 238, 262
   argentea, 72
   compacta, 225

Bolivina, (*contd.*)—
   doniezi, 160, 225
   humilis, 78
   robusta, 59
   vaughani, 225
Bolivinidae, 32, 77—79
Bolivinitidae, 41, 47
Bolivinoides, 72, 89, 90
Bolliella, 37
Botellina, 10
Botellininae, 11, 43
Boultoniinae, 16, 17, 45
Bradyininae, 45
Brizalina, 31, 42
Bronnimannina, 15
Buccella, 33
Bulimina, 20
Buliminacea, 22, 31, 32, 41, 42, 47
Buliminella, 32
Buliminellinae, 32, 47
Buliminellita, 32
Buliminidae, 41, 47
Bulimininae, 47
Buliminoides, 32

C

Calcarinidae, 41, 49
Calcituba polymorpha, 160, 195
Calcivertellinae, 18, 45
Caligellidae, 41, 44
Calvezina, 28
Camerina panamaensis, 131, 132
Candeina, 37
Candeinidae, 36
Candeininae, 35, 48
Candorbulina, 37
Carterina, 39
Carterinacea, 39, 41, 50
Carterinidae, 9, 41, 50
Cassidella, 31
Cassidulinacea, 22, 29, 31, 38, 40—42, 49
Cassidulinidae, 38, 41, 49
Cassigerinella, 37
   chipolensis, 91
Cassigerinellinae, 35, 48
Cassigerinelloita, 37
Catapsydrax dissimilis, 91
   stainforthi, 91
Catapsydracidae, 36, 41, 48
Catapsydracinae, 35, 36
Caucasinidae, 32, 38, 40, 41, 49
Caucasininae, 49

Ceramium, 212, 215
Ceratammina, 10
Ceratobulimina, 31
Ceratobuliminidae, 30, 31, 41, 42, 47
Ceratobulimininae, 30, 31
Chapmanininae, 49
Cheniinae, 16, 17
Chiloguembelina, 37
Chiloguembelinella, 37
Chiloguembelinidae, 31, 37, 41, 48
Chilostomellinae, 39, 49
Chitinosaccus, 11
Chrysalidina, 12
Choffatella, 14
 (*Torinosuella*), 14
Choffatellinae, 14, 43
Chrysalidininae, 9, 12
Chusenellinae, 16, 17, 45
Cibicides, 22
 lobatulus, 160, 183
Cibicididae, 33, 41, 48
Cibicidinae, 48
Cincoriola, 34
Citharina, 27, 72
Citharinella, 27
Claudostriatella, 33
Clavatorella, 37
Clavigerinella, 37
Clavihedbergella, 37
Claviticinella, 37
Clavulinoides, 12
Colania douvillei, 96, 107
Colaniella, 26, 27
 cylindrica, 26
Colaniellidae, 41, 44
Columnella, 13
Colomia, 31
Concavella, 33
Conicarcella, 27
Conoglobigerina, 37
Conorbininae, 29, 32, 33, 47
Conorbitoides, 38
Conorboides, 5, 31
Conorboididae, 31, 32
Conorboidinae, 31
Cornusphaera, 29
Cornuspira lajollaensis, 160
Cornuspirinae, 29
Corrugaella, 33
Coscinophragmatinae, 43
Coskinolina, 14
Coskinolinidae, 9, 14
Coskinolininae, 9, 14, 44

Coskinolinoides, 14
Coxites, 13
Coxitinae, 9, 13
Cribratininae, 43
Cribroeponidinae, 32
Cribrohantkenina, 37
Cribropleurostomella, 38
Cribropleurostomellinae, 38
Cribropullenia, 39
Crithionininae, 8, 11
Cryptomorphina, 28
Cryptoseptida, 28
Ctenorbitoides, 38
Cuneolinella, 15
Cushmanella, 5
Cuvillierininae, 5, 37, 49
Cyclammina, 14, 99
Cyclammininae, 14, 43
Cycloclypeinae, 49
Cycloclypeus, 62—65, 102, 124, 135, 137, 138
 eidae, 63, 65
 postindopacificus var. *tenuitesta*, 63
Cyclogyrinae, 18, 45
Cyclolininae, 44
Cymbaloporidae, 9, 41, 48
Cystamminella, 12
Cystamminellidae, 9, 12

D

Dainella, 16
Dainita, 40
Dainitella, 27
Deckerellina, 31
Delosinidae, 38, 39, 41, 49
Dendronina, 8
Dendrophryinae, 9, 43
Dendrophrynidae, 8, 9, 11
Dendrophryninae, 43
Dentalina, 24, 27
Dictyoconinae, 9, 14, 15, 44
Dictyoconus, 15
 D. (*Paleodictyoconus*), 15
Dicyclinidae, 41, 44
Dicyclininae, 44
Diffusilininae, 11, 43
Diplotremina, 30, 33
Discobotellina biperforata, 160
Discocyclina, 38
Discocyclinidae, 33, 38, 41, 49
Discorbacea, 4, 22, 32, 33, 38, 41, 42, 47
Discorbidae, 32, 33, 41, 47
Discorbinae, 29, 30, 33, 47

Discorbinella, 33
Discorbinopsis aguayoi, 225, 226
Discorbitura, 33
Discospirininae, 46
Dorothia, 13
Dorothiinae, 9, 12, 13
Duostomina, 33
Duostominacea, 29, 30, 41, 42, 47
Duostominidae, 29, 30, 33, 41, 47

E

Earlandiinae, 44
Earlmyersia, 33
Echinoporina, 29
Edithaella, 29
Edithaellinae, 29, 46
Elenia, 15
Ellipsoglandulina, 40
Ellipsopolymorphina, 40
Elphidiidae, 5, 22, 37, 41, 49
Elphidiinae, 5, 49
Elphidium, 22, 37, 89, 99, 121, 215
   advenum, 213
   clavatum, 121, 214—216
   crispum, 160, 182, 226, 231—233
   galvestonense, 213
   gunteri, 213
   incertum, 121, 213—216, 223, 233, 253, 254
   subarcticum, 121, 213, 216
   translucens, 214, 216
Endothyracea, 6, 9, 15, 19, 24, 27, 30, 31, 40, 41, 44
Endothyranopsidae, 15
Endothyranopsinae, 45
Endothyridae, 41, 45
Endothyrininae, 16, 45
Eoendothyra, 16
Eoeponidella, 32
Eofusulininae, 16, 45
Eohastigerinella, 37
Eolagena, 15
Eomarssonella, 13
Eoquasiendothyra, 16
Eorupertia, 34
Eostaffella, 16
Eostaffellidae, 16
Eostaffellina, 16
Eouvigerinidae, 41, 47
Eovolutina, 15
Epistomaria, 5, 8
Epistomariidae, 5, 33, 41, 48

Epistomaroides, 32, 33
Epistominella, 33
Epistominellinae, 32
Epistominidae, 41, 42, 47
Epistomininae, 30, 31
Eponidella, 33
Eponidellinae, 32
Eponididae, 32, 34, 41, 47
Eponidopsidae, 32
Everticyclammina, 14

F

Fabulariinae, 46
Falsopalmula, 27, 28
Fastigiella, 33
Faujasina, 5
Faujasininae, 5, 49
Feurtillia, 14
Fischerinidae, 18, 41, 45
Fischerininae, 45
Fissurina, 40
Flagrina, 9
Flagrinidae, 8, 9, 11
Flectospira, 18
Fourstonella, 15
Frondina, 28
Frondicularia, 25, 27
   woodwardi, 28
Frondinodosaria, 27
Fursenkoina punctata, 69
Fursenkoininae, 49
Fusulinacea, 6, 16, 17, 40, 41, 45
Fusulinellinae, 16, 17, 45
Fusulinidae, 16, 17, 41, 45
Fusulinina, 6, 10, 15, 44
Fusulininae, 16, 45

G

Garantella, 31
Gaudryina, 8, 12, 81, 105
   compacta, 81
   dividens, 81
Gavelinella, 84
   clementina, 70
   lorneiana, 70
Gavelinopsis involutiniformis, 70
   voltziana, 70
Geinitzina, 28
   caseyi, 28
Geinitzinita, 25, 28

Giralarella, 11
Glabratella, 33
  sulcata, 160
Glabratellina, 33
Glandulinidae, 26, 40, 41, 46
Glandulininae, 37
Globanomalina, 37
Globigerapsis, 84
  index, 83
Globigerina, 30, 36, 37, 72, 106
  ampliapertura, 91
  atlantisae, 87
  bulloides, 160, 233, 245
  calida calida, 129
  ciperoensis ciperoensis, 91
  eggeri, 72
  eocaena, 83, 85
  incompta, 87
  pachyderma, 88, 92
  yeguaensis, 75, 84
Globigerinacea, 22, 23, 33—36, 41, 42
  48
Globigerinatella, 37
  insueta, 91
Globigerinatheca, 37
Globigerinelloides, 37
Globigerinidae, 35—37, 41, 42, 48, 75,
  88
Globigerininae, 36, 37, 48
Globigerinita, 37
  glutinata, 245
Globigerinitinae, 35
Globigerinoides, 36, 37, 85, 106, 123
  conglobatus, 36, 245
  pseudobulloides, 84
  ruber, 68, 69, 91, 245
  sacculifer, 36
  trilobus, 75, 84, 85, 120
    —*bisphericus*, 143
    —*primordius*, 143
Globigerinoidesella, 37
Globigerinopsis, 37
Globoconusa daubjergensis, 88
Globoquadrina, 36, 37, 91
  altispira, 91
    —*altispira*, 91
  dehiscens, 5
  dutertrei, 36
  euapertura zone, 5
  praedehiscens, 5
Globorotalia, 19, 34, 36, 37, 74, 85, 91, 108

Globorotalia, (*contd.*)—
  acostaensis, 91
  archaeomenardii/*praemenardii*, 91
  crassiformis, 74, 91, 134
  dutertrei, 91
  fohsi, 85, 87, 91
    —*barisanensis*, 91
    —*fohsi*, 91
    —*lobata*, 91
    —*peripheroronda*, 86
    —*robusta*, 91
  inflata, 91
  kugleri, 91
  lenguaensis, 91
  margaritae, 91, 129
  mayeri, 91
  menardii, 82, 89, 91
  miozea, 115
    —*miozea*, 83
  multicamerata, 88
  opima opima, 91
  pachyderma, 92
  praemenardii, 115
  pseudobulloides, 82, 84, 92, 116
  pseudomenardii, 115
  scitula, 91
  trilobus, 85, 120
  truncatulinoides, 82, 91, 233
  tumida, 91
Globorotaliidae, 35—37, 41, 48
Globorotaliinae, 35—37, 48
Globorotaloides, 36
Globotextulariinae, 13, 44
Globotruncana, 35, 37
Globotruncanella, 35
Globotruncanellinae, 35, 48
Globotruncanidae, 35, 37, 38, 41, 48
Globotruncaninae, 35, 37, 48
Globotruncanita, 35
Globuligerina, 37
Gravellina, 13
Grillita, 29
Gromia, 157
  dujardinii, 157
  oviformis, 158
Gromiida, 7
Gubkinella, 30, 37
Gublerina, 37
Guembelitriella, 37
Guembelitriidae, 30, 35, 37
Guembelitriinae, 48
Guembelitrioides, 37
Guppyella, 15

H

Haeuslerella, 103
Halyphysema, 8
   tumanowiczii, 160
Hantkenina, 37
   (*Bolliella*), 37
   (*Clavigerinella*), 37
Hantkeninidae, 35, 41, 48
Hantkenininae, 35, 37, 48
Haplophragmellinae, 45
Haplophragmoides, 11
Haplophragmoidinae, 43
Harena, 13
Hastigerina, 37
   micra, 91
Hastigerinella, 37
Hastigerininae, 36, 48
Hastigerinoides, 37
Hedbergella, 35, 37
Hedbergellidae, 35, 38, 41, 48
Hedbergellinae, 30, 35
Helenina, 33
Helicolepidininae, 49
Helicorbitoides, 38
Heliozoa, 209
Hemicyclammina, 14
Hemicyclammininae, 9, 14, 43
Hemigordiopsidae, 18, 41, 45
Hemigordiopsis, 18
Hemigordius, 18
Heminwayina, 33
Heminwayininae, 32
Hemisphaerammina, 11
Hemisphaeramminae, 9, 11, 43
Hensonia, 15
Heronallenia, 33
Heronallenita, 33
Heterohelicidae, 31, 35, 37, 41, 48
Heterohelicinae, 35, 48
Heterohelix, 37
Heterolepa, 20—23, 39
Heterolepidae, 33, 39
Heterostegina, 138, 246
   depressa, 160, 246, 251, 252
Hexagonocyclina, 38
Hippocrepina, 11
Hippocrepinella, 11
   alba, 160
Hippocrepinidae, 8, 11
Hippocrepininae, 40, 43
Hirudina, 10
Historbitoides, 38
Homotrematidae, 41, 48

Homotrematinae, 48
Hormosina, 8, 11
Hormosinidae, 9, 11, 40, 41, 43
Hormosininae, 9, 43
Howchinella, 28
Hyperammina, 8, 10, 11
   nodosa, 8
Hyperamminidae, 9
Hyperammininae, 10, 11

I

Ichthyolaria, 24—27
Indicola, 33, 34
Indicolidae, 33
Inordinatosphaera, 37
Involutina, 29, 39
Involutinidae, 16, 29, 39, 41, 47
Iridia diaphana, 160
   lucida, 160
Iraqia, 15
Irregularina, 10
Islandiellidae, 41, 47
Iuliusina, 35

J

Jadammina macrescens, 213
Jullienella, 68

K

Kahlerininae, 16, 17
Keramosphaerinae, 46
Kilianina, 14
Klubovella, 16
Kollmannita, 30
Kurnubia, 15
Kurnubiinae, 9, 14, 44
Kyphamminidae, 10

L

Lacosteininae, 47
Lagenumbella, 15
Lagynacea, 41, 42
Lagynidae, 7, 11, 41, 42
Lagyninae, 7
Langella, 28
Lasiodiscidae, 41, 45
Laticarinina, 33
Laticarininidae, 31, 41, 48
Lenticulina, 24, 26
Lenticulinidae, 26, 27
Lenticulininae, 27, 46

Lepidocyclina, 60—62, 108, 135, 138
  dartoni, 61
  howchini, 61
  isolepidinoides, 61
  japonica, 61
  parva, 61
  peruviana, 61
  pustulosa, 61
  radiata, 60
  yurnagunensis, 61
Lepidocyclinidae, 33, 38, 41, 49
Lepidocyclininae, 49
Lepidorbitoides, 38, 106, 108, 136
Lepidorbitoididae, 38, 41, 49
Lingulina costata, 28
      —*tricarinata*, 28
  infirmis, 28
Lingulinella, 27
Lingulininae, 26, 46
Lingulonodosaria, 27
Lituolacea, 8, 9, 13, 19, 30, 40, 41, 43
Lituolidae, 9, 11, 13, 41, 43
Lituolidea, 9
Lituolinae, 43
Lituonella, 14
Lituotubellinae, 16
Loeblichella, 35
Loeblichellinae, 35
Loeblichia, 16, 17
  (*Urbanella*), 16
Loeblichiidae, 16, 17, 27, 40, 41, 44
Loeblichiinae, 16
Loeblichinidae, 16
Loeblichopsis, 8, 11
Loftusiidae, 9, 14, 41, 43
Loftusiinae, 14, 43
Loxostomatidae, 38, 41, 49
Lunatriella, 37
Lunacammina, 24, 27, 28

M

Madreporites lenticularis, 15
Marginopora, 252
  vertebralis, 160, 246
Marginotruncana, 35
Marginotruncanidae, 35
Marginulina, 27
Marginulininae, 27, 46
Marginulinopsis, 27
Marieita, 15
Marsipellinae, 8, 11

Marssonella, 13
  oxycona, 62
Martinottiella, 13
Martiguesia, 14
Martyschiella, 12
Massilina, 225
Mayncella, 14
Meandropsininae, 46
Meandrospira, 18
Mediocris, 16
Megastomella, 33
Melonis, 20, 21
  scaphum, 21
Merlingina, 13
Mesodentalina, 24, 26, 27
Mesoendothyrinae, 14, 43
Metarotaliella parva, 160
Meyendorffina, 14, 15
  (*Paracoskinolina*), 15
Migros, 12
Miliammina fusca, 216
Miliolacea, 13, 19, 41, 45
Miliolidae, 18, 41, 46, 157
Miliolina, 6, 7, 17—19, 45
Miliolinae, 46
Miliolinella subrotunda, 17
Miliolinellinae, 46
Miliolipora, 18
Milioliporidae, 17, 41, 46
Millettia, 31, 32
Minouxia, 12
Miogypsina, 62, 67, 135
  tani, 66
Miogypsinidae, 41, 49, 67
Miogypsinoides, 108
Miscellaneidae, 37
Misellininae, 16, 17, 45
Monodiexodina, 95
  kattaensis, 94
Moravamminidae, 41, 44
Moravammininae, 44
Multiseptida, 26
Myxotheca arenilega, 160
Myxothecinae, 7

N

Nankinellinae, 17
Nemogullmia longevariabilis, 160
Neocribrella, 34
Neodiscocyclina, 38
Neoflabellina, 26
Neoglabratella, 33

Neoiraqia, 15
Neoschwagerinidae, 16, 17, 41, 45
Neoschwagerininae, 16, 17, 45
Neothailandina, 17
Neotuberitinae, 16
Neusina, 9
Neusinidae, 9
Nezzazata, 13
Nezzazatidae, 9, 13, 19
Nezzazatinae, 13
Nipterula, 27
Nodobaculariinae, 46
Nodosarella, 40
Nodosaria, 20, 24, 26, 27
   affinis, 25
Nodosariacea, 6, 22, 24—27, 30, 40—42,
   46
Nodosariidae, 16, 24—27, 31, 40—42, 46
Nodosariinae, 26, 27, 46
Nodosinella, 27, 28
   digitata, 28
Nodosinellidae, 24, 25, 27, 31, 40—42, 44
Nodosinellinae, 40, 44
Nonionacea, 4, 5, 22, 39, 40—42, 49
Nonionella, 88
Nonionidae, 5, 22, 37, 39, 41, 49
Nonioninae, 49
Nouriidae, 41, 43
Novatrix, 13
Nubeculariidae, 18, 41, 46
Nubeculariinae, 46
Nummulites, 97, 98
   aquitanicus, 97
   exilis, 97
      —involutus, 97
      —planulatus, 97
      —robustus, 97
   jacquoti, 97
   laevigatus, 97
   nitidus, 97
   panamaensis, 97
   planulatus cussacensis, 97
   praelaevigatus, 97
   thaliciformis, 97
   vonderschmitti, 97
Nummulitidae, 5, 49, 97, 98, 108
Nummulitinae, 49

O

Oberhauserella, 30
Oberhauserellidae, 30, 41, 42, 47
Oolininae, 46

Operculina, 60
Opertum, 13
Ophthalmidiinae, 46
Orbignyana, 13
   (*Ataxoorbignyana*), 13
Orbiqia, 14
Orbitammina, 14
Orbitoclypeidae, 38
Orbitoclypeus, 38
Orbitocyclina, 38
Orbitocyclinoides, 38
Orbitoidacea, 4, 22, 33, 38, 41, 49
Orbitoididae, 33, 38, 41, 49
Orbitolina, 15
   O. (*Praeortitolina*), 15
Orbitolinidae, 9, 13—15, 41, 44
Orbitolininae, 44
Orbitolinopsis, 15
Orbitolites, 252
   duplex, 245
Orbulina, 37, 143, 238
Orbulininae, 36
Orbulinoides, 37
Orbulites lenticulata, 15
Ordovicina, 10
Osangulariidae, 39, 41, 49
Ovalveolina, 13, 19
Oxinoxisidae, 8, 11, 41, 43
Ozawainellidae, 16, 17, 41, 45
Ozawainellinae, 17, 45

P

Pachyphloia, 24, 27
Pachyphloides, 28
   oberhauseri, 28
Padangia, 28
Palaeonubecularia, 15
Palaeotextulariidae, 9, 31, 41, 44
Paleodictyoconus, 15
Palmerinella, 33
Palmula, 26, 28
Palorbitolina, 15
Paracoskinolina, 15
Paracyclammina, 14
Parafissurina, 40
Paragaudryina, 12
Paralingulina, 28
Paramillerella, 16
Parabulloides, 16
Pararotalia, 37
Pararotaliinae, 37
Paraschwagerina, 96

Parathurammina, 10
Parathuramminacea, 10, 15, 41, 44
Parathuramminidae, 10, 41, 44
Paratrocholina, 29
Pasternakia, 13
Patellina corrugata, 160, 225
Patellinella, 29
Patellininae, 29, 47
Pavonitina, 14
Pavonitinidae, 14
Pavonitininae, 14, 47
Pegidiidae, 33, 41, 47
Pellatispira, 111
   glabra, 112
Pelosphaera cornuta, 8
Peneroplidae, 5
Peneroplinae, 46
Peneroplis, 18, 246, 252
Permodiscus, 29
Pernerina, 13
Pfenderina, 14
Pfenderella, 14
Pfenderinidae, 13, 14, 41, 44
Pfenderininae, 14, 44
Phyllopsammia, 14
Pijpersia, 33
Pileolina, 33
Pilulina, 11
   jeffreysi, 7, 8
Pilulininae, 11
Pinaria, 39
Placopsilinidae, 9, 11
Placopsilininae, 11, 43
Plagiostomella, 30, 39
Planoendothyra, 16
Planoglabratella, 33
Planomalina, 37
Planomalinidae, 35, 41, 48
Planomalininae, 35, 37
Planorbulina, 183, 264
   mediterranensis, 160, 183, 264
   vulgaris, 264
Planorbulinella, 62, 134
   larvata, 135
   zelandica, 135
Planorbulinidae, 41, 48
Planularia, 27
Planulariinae, 27
Planulina, 20, 23
Planulinacea, 33
Planulinidae, 33
Planulininae, 48
Planulinoides, 33

Plectofrondiculariinae, 26, 46
Plectorecurvoidinae, 44
Pleurostomella jurassica, 40
Pleurostomellidae, 38, 40, 41, 49
Pleurostomellinae, 49
Plummerita, 37
Polydiexodininae, 16, 17, 45
Polymorphina, 20
Polymorphinidae, 26, 29, 41, 46
Polymorphininae, 29, 46
Polysiphonia, 212, 215
Polystomella strigilata, 157
Praebulimina, 32
Praeglobotruncana, 37, 82, 83
   delrioensis, 83
   marginaculeata, 83
   stephani, 83
     —var. *gibba*, 83
Praegubkinella, 30, 42
Praeindicola, 34
Praekurnubia, 15
Praeorbitolina, 15
Praeorbulina, 143
   glomerosa, 91
Praereticulinella, 13
Pravoslavlevia, 27
Prodentalina, 24—27
Proporocyclina, 38
Protelphidium tisburyensis, 213—216
Protobotellina, 8
   cylindrica, 7, 8
Protonodosaria, 28, 29
Protopeneroplis, 16
Protumbella, 15
Psammosphaera, 8
Psammosphaeridae, 10
Psammosphaerinae, 10, 11, 43
Pseudoarcella, 27
Pseudobolivininae, 44
Pseudochoffatella, 14
Pseudocyclammina, 14, 99
Pseudodoliolininae, 16, 17, 45
Pseudoendothyra, 16
Pseudoendothyridae, 16, 17
Pseudoendothyrinae, 45
Pseudofrondicularia, 28
Pseudofusulininae, 17
Pseudogaudryina, 12
Pseudoguembelina, 37
Pseudohastigerina, 99
Pseudolangella, 28
Pseudomarssonella, 13
Pseudonodosaria, 26—28

Pseudoparrella, 33
Pseudoparrellidae, 32, 33, 41, 48
Pseudophragmina, 38
Pseudopyrulinoides, 29
Pseudorbitoides, 38
Pseudorbitoididae, 33, 38, 41, 49
Pseudorbitoidinae, 38, 49
Pseudoreophacinae, 8
Pseudoreophax, 8
Pseudoruttenia, 33
Pseudoschwagerina, 96, 127
Pseudoschwagerininae, 16, 45
Pseudostaffellinae, 17, 45
Pseudotextulariella, 15
Pseudotriplasia, 15
Pseudoumbella, 15
Pseudowebbinella, 11
Ptychocladiidae, 15, 41, 44
Pullenia, 39
Pulleniatina, 19, 34, 36, 37, 91
  obliquiloculata, 91
Pullenidae, 39

Q

Quadratobuliminella, 32
Quasiendothyra, 16
  (Eoendothyra), 16
  (Eoquasiendothyra), 16
  (Klubovella), 16
Quasiendothyridae, 16, 17
Quasiumbella, 15
Quasiumbelloides, 15
Quinqueloculina, 17, 18, 214, 215, 233
  lata, 213, 221, 222, 225, 238
  seminulum, 17, 213, 214, 216
Quinqueloculininae, 46

R

Rabanitina, 13
Radotruncana, 37
Raibosammina, 10
Rakusia, 29
Ramulininae, 46
Rectocyclammina, 14
Rectodictyoconus, 15
Reinholdella, 31
Reinholdellinae, 31
Remananeicinae, 44
Remanellina, 27
Reophacidae, 9
Reophacinae, 9, 10, 43

Reophax, 8
Reticulinella, 13
Reticulophragmium, 14
Reticulopodia, 7
Rhabdammina, 8
Rhabdammininae, 10
Rhabdorbitoides, 38
Rhapydionininae, 46
Rhizammininae, 9
Rhizammininae, 10, 42
Rhynchogromiinae, 7
Riyadhella, 13
Robertina, 5
Robertinacea, 23, 30, 41, 42, 47
Robertinidae, 29—31, 40—42, 47
Robertininae, 30, 47
Robuloides, 16
Robuloidinae, 15, 16, 45
Robulus, 26
Rosalina, 32, 33, 69, 224, 232
  columbiensis, 225
  floridana, 69, 160, 230, 231
  hyalina, 243
  leei, 160, 213, 216, 221, 222, 224—227,
    232—235, 237—240, 244, 251
  vilardeboana, 161
Rosalinidae, 32, 33
Rosalininae, 32, 47
Rotalia, 23, 157
Rotaliacea, 22, 23, 37, 41, 48
Rotaliella, 222
  heterocaryotica, 161, 195, 199, 225, 226,
    232
  roscoffensis, 161, 195, 225, 232
Rotaliellidae, 41, 47
Rotaliidae, 5, 30, 33, 41, 48, 157
Rotaliina, 6, 9, 19, 21, 46
Rotaliinae, 5, 37, 49
Rotalina nitida, 157
Rotalipora, 37, 82
Rotaliporidae, 35
Rotaliporinae, 35, 37, 48
Rubratella, 31
Rugoglobigerina, 35, 37
Rugoglobigerininae, 35
Rugotruncana, 35, 37
Rupertininae, 33, 48
Rzehakinidae, 9, 40, 41, 43

S

Saccammina, 8, 11
  alba, 161

Saccamminidae, 9, 10, 41, 43
Saccammininae, 10, 11, 43
Saccamminoides, 10
Saccamminopsis, 10
Saccorhiza, 11
Sagenina, 9
Sagrina, 31
Sanderella, 14
Saracenaria, 27
Schackoina, 37
   S. (*Eohastigerinella*), 37
   S. (*Hastigerinoides*), 37
Schackoinella, 37
   S. (Beella), 37
Schackoinidae, 35, 41, 48
Schenckiella, 13
Schizammina, 68
Schizamminidae, 41, 43
Schlagerina, 30
Schmidtia, 30
Schubertellidae, 17, 41, 45
Schubertellinae, 16, 17, 45
Schubertia, 32
Schubertiinae, 32
Schwagerina, 96
Schwagerinidae, 41, 45
Schwagerininae, 16, 17, 45
Seabrookiina, 46
Semitextulariidae, 41, 44
Serovaininae, 33, 47
Serpenulina, 11
Shepheardella, 7, 179
   taeniformis, 7, 161
Shidelerella, 10
Siderolites, 108
Sieberina, 27
Silicammina, 10
Silicamminidae, 8, 10, 11
Silicotuba, 10
Silicotubidae, 8, 10, 11
Siphogaudryina, 12
Siphogenerinoides, 31
Siphoninidae, 33, 41, 48
Sorites, 252
Soritidae, 18, 41, 46
Soritinae, 46
Sorosphaera, 11
Sosninella, 28
Sphaerammininae, 43
Sphaeroidinella, 34, 37
   dehiscens, 36
Sphaeroidinellopsis, 37
Sphaeroidinidae, 41, 47

Spinarcella, 27
Spinosphenia, 27
Spirillina vivipara, 161, 225
Spirillinacea, 6, 29, 39, 41, 47
Spirillinidae, 41, 47
Spirillininae, 29, 47
Spirocyclinidae, 14
Spirocyclininae, 14, 43
Spirolina, 252
Spiroloculina hyalina, 157, 161, 221, 222, 224—228, 230, 231, 233, 235, 237—242, 244
Spiroloculininae, 18, 46
Spiroplectammina olokunsui, 116
Spiroplectammininae, 43
Spiroplectinata, 12, 80, 105
   lata, 81
Squamulinidae, 18, 41, 46
Stacheia, 15
Stacheiinae, 15
Stacheioides, 15
Staffellidae, 16, 17, 41, 45
Staffellinae, 16, 45
Stainforthiinae, 32
Stannophyllum, 9
Stegnammina, 10
Stegnamminidae, 10, 41, 43
Stegnammininae, 10
Steinekella, 14
Stenocyclina, 38
Stensioina, 71
   altissima, 71
   exsculpta, 71
      —*aspera*, 71
      —(mut.) *infirma*, 71
   gracilis, 71
   pommerana, 70, 71
      —var. *juvenilis*, 71
   prae-exsculpta, 71
      —var. *granulata*, 71
Stetsonia, 33
Streblus (=*Rotalia*) *beccarii tepida*, 161
Streptocyclammina, 14
Stomatorbina, 31
Sulcorbitoides, 38
Sumatrininae, 16, 17, 45

T

Tauridia, 28
Tawitawiinae, 14, 44
Technitella, 11
Tetrataxidae, 9, 41, 44
Textularia, 79

Texturariella, 15
Textulariellidae, 9, 15, 44
Textulariidae, 9, 14, 31, 41, 43
Textulariina, 6, 7, 42
Textulariinae, 43
Thailandina, 17
Thailandininae, 17, 45
Thekammina, 10
Tholosininae, 8, 9, 11
Thurammina, 10
Thuramminidae, 10
Thuramminidae, 8, 10
Ticinella, 37
Tinogullmia, 7
Tollmannia, 28
Tolypammina, 11
Tolypamminidae, 9
Tolypammininae, 43
Torinosuella, 14
Torresina, 33
Tournayellidae, 41, 44
Tremachoridae, 39, 41, 49
Tretomphalus, 32, 225
Triasina, 29
Trichohyalus, 33
Trifarininae, 32
Triloculina, 18
Tritaxia, 12
Trochammina, 12
   depressa, 89
   inflata, 213—216, 238, 254
Trochamminidae, 9, 10, 12, 41, 44
Trochammininae, 44
Trocholina, 29
Trocholinidae, 29
Trochoporina, 12
   globigeriniformis, 12
Trochoporininae, 9, 12
Trochospira, 13
Truncorotaliinae, 37
Truncorotaloides, 37
Truncorotaloidinae, 36, 37
Tscherdyncevella, 15
Tubertininae, 15, 44
Tubinellinae, 46
Turborotalia, 36, 37
Turrilina, 21, 22
   brevispira, 21
Turrilinidae, 41, 47
Turrilininae, 47

U

Umbellina, 15

Umbellinaceae, 15
Umbellininae, 15
Ungdarellaceae, 15
Ungulatella, 29
Ungulatellinae, 29, 47
Ungulatelloides, 29
Uralinella, 10
Urbanella, 16
Urnulella, 27
Usbekistania, 11
Usbekistaniidae, 11, 12
Uslonia, 10
Uslonianinae, 10
Usloniinae, 8, 10
Uvigerina, 31, 79, 80
   cretensis, 132
Uvigerinidae, 32, 41, 47

V

Vaginulina, 27
Vaginulinidae, 26, 27, 40, 41, 46
Vaginulininae, 27, 46
Vaginulinopsinae, 27
Vaginulinopsis, 27
Valvoreussella, 12
Valvulina, 12
Valvulininae, 12, 13, 44
Variostoma, 33
Variostomatidae, 30
Vaughanina, 38
Vaughanininae, 38, 49
Vellaena, 31
Ventilabrella, 37
Verbeekinidae, 16, 17, 41, 45
Verbeekininae, 16, 45
Verneuilinella, 13
Verneuilinidae, 9
Verneuilininae, 12, 44
Vernonina, 33
Vialovella, 13
Victoriellinae, 34, 48
Virgulinidae, 32

W

Webbinellinae, 29, 46
Wedekindellininae, 16, 17
Weikkoella, 10
Wheelerellinae, 49
Whiteinella, 35, 37

Y

Yvonniellina, 27